100
MILITARY
INVENTIONS

CHANGED
THE WORLD

100 MILITARY INVENTIONS THAT CHANGED THE WORLD

ROD GREEN

A Herman Graf book
Skyhorse Publishing

Skyhorse Publishing books may be purchased in bulk
at special discounts for sales promotion, corporate gifts,
fund-raising, or educational purposes.
Special editions can also be created to specifications.
For details, contact the Special Sales Department,
Skyhorse Publishing, 307 West 36th Street,
11th Floor, New York, NY 10018
or info@skyhorsepublishing.com.

www.skyhorsepublishing.com

ISBN 978 1 62087 563 6

Printed and bound by CPI Group (UK) Ltd, Croydon, CR0 4YY

10 9 8 7 6 5 4 3 2 1

Library of Congress Cataloging-in-Publication Data available on file

CONTENTS

INTRODUCTION

100 MILITARY INVENTIONS THAT CHANGED THE WORLD

Deciding exactly what to include in a book of military inventions requires a bit of head scratching. What precisely constitutes a 'military invention'? Not everything used by the military was purposefully invented for them, after all. Over the centuries, soldiers, sailors and airmen have been issued with mountains of kit to keep them healthy, well fed, warm and dry – from toothpaste to tarpaulins – but not everything that comes as part of a soldier's kit was necessarily originally designed by or for the military. Neither can many of the things that service personnel need to do their jobs be described as items that have 'changed the world', which is the other major criterion for inclusion in our list.

Some things, like the hand grenade or the Kalashnikov assault rifle are clearly intended for military use and have had such an influence on military planning, battlefield tactics or public perception that they can be said to have changed the world. Some things, like tanks, aircraft carriers or the atomic bomb, are obvious military items that have most definitely changed the world. They can be given a big, confident 'tick' and slotted into the book.

Yet there are other inventions that might not at first appear to have any military connection that have influenced the world around us in an extraordinary variety of ways. Preserving food by sealing it in tin cans is a technique that was developed primarily because there was a military market for the end product. Titanium is in widespread use in the modern world only because a metal that was incredibly strong, yet lightweight and resistant to corrosion, had obvious military applications.

And then there is the space programme. Over the years, NASA (National Aeronautics and Space Administration) has cultivated an image as a civilian scientific organisation and its first director, Dr Thomas Glennan, was an engineer and academic. But the organisations that Dr Glennan brought together under the NASA umbrella included the US Air Force and the Department of Defense's Advanced Research Projects Agency, the US Army Ballistic Missile Agency and part of the Naval Research Laboratory. The first astronaut, Alan Shepard, was a former US Navy fighter pilot; the first man on the Moon, Neil Armstrong, was also a former US Navy fighter pilot; and the space race was, in any case, inspired by the Cold War and a desperate rush to put spy satellites into orbit that could photograph military installations. Anything, therefore, that was invented by or in association with NASA, or was developed as a spin-off from the US space programme, can surely be regarded as fair game and should be included in this book ... as long as it changed the world a bit.

Space-age inventions constitute a fair proportion of the book, but do not dominate the pages completely. There are many ancient devices, like the trebuchet or the cannon, that can be said to have changed the course of battles, altered the political map and, therefore, changed the world. The Chinese crop up on a regular basis, having invented gunpowder and discovered a host of other fascinating things, such as the principles of helicopter flight. They didn't always develop their discoveries into military inventions but they often did most of the ground work centuries before the more industrialised nations of the western world turned Chinese ideas into weapons of warfare.

Because so many of the inventions featured have a history that can be traced back hundreds, if not thousands of years, the idea of putting them in any kind of chronological order quickly had to be abandoned. In the end, the entries work far better presented in a more random way, allowing you to 'dip in' to the text without always finding the shiny stainless steel and titanium things at the end of the book and the creaky old wooden stuff at the beginning. Hopefully, that helps to make the whole book as entertaining as possible, with a few surprise inventions that you never realised had anything whatsoever to do with the military cropping up amongst the military hardware of death and destruction.

DINNER
IN A TIN

(CANNED FOOD)

An army marches on its stomach, so the saying goes, but feeding an army, or sailors who might have to spend months on end at sea, has always posed something of a problem. HMS *Victory*, famously Admiral Lord Nelson's flagship at the Battle of Trafalgar in 1805, carried enough provisions to sustain her crew of 821 for up to six months. Pork and beef were heavily salted and packed in barrels, and there were also barrels of fresh water, peas, vinegar, wine and beer. The barrels were heavy, making them good for ballast on board ship but very difficult for an army to transport cross country. In any case, the contents of the wooden barrels would inevitably spoil after a time. Ships had to head for port to be resupplied and armies had to scour the countryside for food.

Until Louis Pasteur's experiments in the 1860s proved that micro-organisms were responsible for food going off, no one understood exactly why packing things in barrels was a futile exercise without first having sterilised the containers and their contents to kill the bacteria and then sealing them airtight to keep out all the micro-bugs. People simply knew that cramming food into barrels, even when the food was heavily salted, wouldn't preserve it indefinitely. It was a major headache for military planners. They had vast armies to feed and when thousands of hungry soldiers arrived in an area they tended to strip the place of food like a plague of locusts. This filled everyone with dread when they heard the army was coming to town.

To try to find a solution, in 1795 the French military offered a substantial reward of 12,000 francs for anyone who could come up with a successful way of preserving food in a manner that could easily be transported. A French chef in Paris, Nicolas Appert was already experimenting with ways of preserving food. In the days before refrigerators, after all, food storage was a problem that affected everyone, not just the military. It took him 15 years, but Appert eventually developed a solution that seemed to work. He put food into thick glass bottles and pushed a cork tightly into the neck of the bottle, leaving a little space for air between the cork and the contents. He then wrapped the bottles in canvas to protect them, and immersed them in boiling water until he reckoned the contents were cooked. The corks were further sealed with wax and the food inside was safely preserved. In 1810 Appert was awarded the 12,000-franc prize and published a book about preserving food.

Appert's method was sound, but the end product was far from perfect. Heavy glass jars were expensive and liable to get broken when being transported in bulk to feed an army. Something more lightweight was required and an Englishman named Peter Durand expanded on Appert's technique, filing a patent in 1810 for preserving food in tin cylinders, or canisters, a word that was soon shortened to 'cans'. Durand sold his patent to businessmen Bryan Donkin and John Hall. As Donkin, Hall and Gamble, their business was eventually to become part of the food manufacturer Crosse and Blackwell.

The new cans were made from tin-plated wrought iron. The wall of the can was made from a flat rectangle of metal that was rolled into a cylinder shape and then soldered down the seam. A disc of metal was then soldered to make the bottom and the food was inserted, sometimes partly cooked, before the lid was soldered on with a gap left to allow steam to escape as the cooking process was completed. The gap was then soldered shut to seal the can.

The technique worked, although the lead solder used at the time would undoubtedly have been highly toxic. Donkin sold canned meat to the British Army and Royal Navy, with Arctic explorer Sir John Franklin taking canned food on his Arctic expedition to establish the North-West Passage in 1845. Franklin and his crew disappeared on that mission but some of his stores were found by a subsequent expedition. One of the cans was opened almost a century later and the food found still to be edible.

For a time, canned food was fashionable with wealthy diners, being seen as a novel and suitably expensive way to eat for those who had money to squander, but as the canning process became more advanced, more automated and cheaper, canned goods soon became more affordable to buy and hugely popular. The biggest innovation was in the process of making the can. The time-consuming and costly lead solder was made redundant by the beginning of the 20th century when rolling and folding processes were devised that folded and squeezed the open ends of the cylinder wall together to make it airtight. Crimping, folding and rolling the ends of the can under great pressure also created airtight seals for the top and bottom of the can.

By the time of the First World War in 1914, it was possible to issue soldiers with individual rations in cans and almost anything could be supplied canned – except beer. Beer proved a little problematic due to its gassy nature but in 1935 the first canned beer went on sale, although you still needed some kind of can opener to get at it. A range of clever, hand-held, blade-and-lever can openers had been devised, as using the soldiers' traditional method, slicing the can open with your bayonet, was liable to lose you a finger. And no one but a soldier is likely to have a bayonet at the ready when he fancies a beer. If you decided to wait for a ring pull you'd have been pretty thirsty. That didn't come along until 1959.

SHARPE'S CHOICE
AT ANY RATE

(THE BOLT ACTION RIFLE)

Fans of Bernard Cornwell's fictional military hero Richard Sharpe will have read about and seen him (on TV as played by actor Sean Bean) drilling squads of 19th-century soldiers to load and fire their muskets at a rate of three rounds per minute. Sharpe himself was proficient enough to be able to manage five with a musket, less with the Baker rifle that he generally carried. The whole procedure was similar to loading a cannon.

Gunpowder from a flask was poured down the barrel of the weapon. A lead ball and some wadding, to ensure the ball fitted tightly into the barrel, was then rammed all the way down the barrel using a metal ramrod. The spring-loaded hammer was pulled back to cock the mechanism and a touch of gunpowder was then added to a flash pan striking plate at the breech of the weapon. When the trigger was pulled, the hammer hit a flange of metal above the flash pan, the flint in the hammer causing a spark that ignited the powder there causing a flash. The flash passed through a small hole above the combustion chamber in the barrel, igniting the main gunpowder charge and firing the weapon.

Given that the above is a quick description of the firing procedure, it's not difficult to see why Sharpe's men struggled to fire more than three rounds a minute. Like everything else during the time of the Industrial Revolution, the lot of the infantry soldier was to change dramatically, although Sharpe was long retired before British troops started to receive a new type of weapon. The 1853 Pattern Enfield rifle sped things up a little. The main powder charge and the lead ball came together in a paper cartridge. The soldier bit open the cartridge, poured the powder down the barrel, then rammed home the ball using the paper cartridge as wadding. Pulling the trigger released the hammer onto a percussion cap that caused a spark, doing away with the old flint-and-powder system. The average soldier could now manage four rounds per minute, but there was a problem with the cartridges in India. The paper was lubricated with either pork or beef fat, causing deep offence to the religious sensibilities of both Muslim and Hindu soldiers and contributing to the Indian Mutiny of 1857.

The next big innovation did away with the lengthy procedure of pouring powder down the barrel when metal cartridges that had a percussion cap at the rear, a powder-filled cartridge case in the middle and a cone-shaped bullet at the end were introduced. These were loaded into the breech of the rifle and the trigger operated a hammer that struck the percussion cap, firing the weapon. The soldier then had to open the breech and remove the spent cartridge case. When the British Army adopted this system with the Snider-Enfield rifle in 1867, the infantryman's rate of fire soared to ten rounds per minute. Within a few years, however, things were to change completely.

In 1871 the Martini-Henry rifle came along, using a lever-action system. Like the famous Winchester, it used a lever beneath the stock to load a cartridge from a magazine and cock the weapon. The rifleman then fired using the trigger; and the lever ejected the spent cartridge before loading a new one. The British soldier could now fire a dozen or more rounds every minute, most importantly without having to stand up, move about or even take the rifle butt from his shoulder.

More than half a million Martini-Henry rifles were made and it remained in service with the British Army for almost 20 years but was superseded by the weapon that became the infantryman's best friend – the bolt action rifle. The British squaddies' first experience of a bolt action rifle came with the Lee-Metford in 1888. The bolt action was a major step forward because it was engineered in such a way that it used fewer moving parts than other systems, making it less liable to jam. It was tough, reliable and easy to use.

Lifting the bolt handle and pulling it back ejected a spent cartridge and cocked the weapon. Pushing the bolt forward again loaded a fresh cartridge from the magazine. The principle had been around for more than a century but the engineering technology that allowed for its mass production, and the ammunition that made the weapon viable, came together at the end of the 19th century, providing the rifleman with a weapon that Richard Sharpe would have loved. British infantrymen were expected to be able to hit a target 300 yards away (273 metres) with 15 rounds in a minute. When the Lee-Enfield, a development of the Metford, was introduced in 1895 the rate of fire went even higher. In 1914, British Army small arms instructor Sergeant Snoxall set the record at 38 rounds on target in one minute. When German troops first encountered British units armed with Lee-Enfield rifles during the First World War, they reported that they had come under fire from multiple machine guns.

The Lee-Enfield bolt action rifle, in various guises, became the standard infantry weapon for British and Commonwealth forces, remaining in service with the British Army through two World Wars up to 1957. More than 17 million Lee-Enfields have been produced and it is still in service as a general issue weapon and sniper rifle with military units and police forces around the world.

Because their robust simplicity allows them to use more powerful ammunition and maintain a greater degree of accuracy, bolt action rifles are ideal for hunters and snipers. The Lee-Enfield is far from being the only example of its type, but it is one of the most ubiquitous infantry weapons of all time and, without doubt, a bolt action Lee-Enfield would have been Richard Sharpe's rifle of choice.

MARATHON MAN
ON THE MOON

(CUSHION-SOLED RUNNING SHOES)

When Neil Armstrong became the first man to walk on the Moon in July 1969, two weeks before his 39th birthday, he probably wasn't thinking too much about running a marathon. Given that he was about to become the first human being in history to set foot on a chunk of rock in space that is not our own planet, he was probably more concerned about not stumbling over the words of his 'One small step for a man ...' speech than he was about anything else, although most of the rest of us would surely have been thinking, 'What if I just sink into the surface never to be seen again?'

Armstrong knew better than that. As well as having been a military combat pilot and test pilot, he was a scientist and engineer, so he knew pretty well what to expect on the Moon and was prepared to cope with it. He knew that, after four days of weightlessness in outer space on the journey between Earth and the Moon, he was about to feel heavy again, although the Moon's gravitational pull is much weaker than that on Earth. He would be just one-sixth of his Earth weight, but he was dressed in a multi-layered, pressurized, insulated, temperature-controlled, meteorite-proof space suit that weighed almost 200lb

(91 kg) – far more than he did. The bulky suit limited his movements meaning that, athletic as he was, his body needed to be able to cope with the stresses and strains of bouncing around on the lunar surface.

NASA's space-suit designers fully realised that walking on the Moon would be a strange experience for Armstrong, and his colleague Buzz Aldrin when Aldrin joined him on the lunar surface, and built a special feature into their Moon boots to help absorb impacts, cushion their feet and protect their joints. The specially developed foam layer in the soles of their boots not only protected the astronauts but also helped to ventilate their boots. In the end, the pair claimed to have found walking on the Moon far easier than the simulated Moon walks they had gone through during training, but the efforts of the Moon boot designers had not gone to waste. The cushioned soles performed perfectly and it was an idea that was to endure well beyond the space race.

In 1986, sports shoe company KangaROOS launched a running shoe range featuring a 'Dynacoil' layer in the sole. Dynacoil was a three-dimensional shock absorption system designed to distribute evenly the force of your foot hitting the ground and bounce the energy back to you, putting a spring in your step. Dynacoil was developed with the cooperation of NASA, using the lessons they learned in making Neil Armstrong's Moon boot, and was marketed as an aid to runners and walkers that would help them avoid stress injuries from pounding the pavement, as well as being a boon to other sportsmen such as basketball players, as it gave them more bounce for jumping.

These were not the first sports shoes with space-age cushioning. That honour goes to the Nike Tailwind, the shoe that launched the Nike Air series in 1978. The idea of creating a running shoe that featured a hollow sole air pocket came from former NASA engineer Frank Rudy, who used a technique called 'blow rubber moulding', previously employed to create NASA space helmets!

So, Neil Armstrong may not have been thinking about running a marathon when he stepped onto the Moon, but his feet were almost ready for it!

KITTY LITTER
WORTH A BOMB

(DYNAMITE)

Next time you are clearing out the cat's litter tray, desperately searching for something to think about that will take your mind off the tedious and unpleasant task at hand, something that will make the job go a bit faster, why not try dynamite? No, that's not to suggest that you blow the litter tray to bits – that would be very messy – but your cat litter has more in common with dynamite than you might realise.

Dynamite, of course, was invented by Alfred Nobel. Nobel was born in Stockholm in 1833, the son of Immanuel Nobel, a businessman, builder and engineer with a love of explosives. Immanuel was interested in the use of explosives in civil engineering but, after establishing a factory in Russia in 1837, he began producing munitions, naval mines and even experimented with torpedoes. War is always good for the munitions business and Nobel's factory flourished throughout the Crimean War but, when peace broke out, Nobel's financial situation was one of the first casualties. His creditors insisted that he step down as head of the business, installing Alfred's elder brother Ludvig in his place. Immanuel returned to Sweden where he established a factory to produce the explosive nitroglycerin.

Nitroglycerin is an extremely unstable substance that is prone to explode if it gets too hot, or if it freezes (which it does at 13° Celsius) and starts to thaw out, or if it is jostled – and whatever you do, don't drop it. Handling it is enough to give you a heart condition, but if it does, your doctor may prescribe nitroglycerin tablets, a dilute form of the chemical which has been used for many years to treat heart disease! The Nobel family found out to their cost how dangerous nitroglycerin was when Immanuel's youngest son, Emil, along with several factory workers, was killed in an explosion in 1864. After his son's death, Immanuel suffered from chronic ill health and died six years later. Alfred, however, was determined to find a safe way to manufacture, store and transport nitroglycerin. Because it was so unstable as a liquid, he experimented in combining it with other things. The concoction that seemed to work best was a mixture of sodium carbonate and diatomaceous earth. Diatomaceous earth is a powdery substance formed from the remains of fossilized crusty algae. It has many uses, including, nowadays, in non-clumping kitty litter.

Alfred patented his new explosive in 1867, calling it Dynamite after the Greek word *dynamis*, which means 'power'. The new Dynamite could be rolled into sticks and wrapped in paper to protect it. It was perfectly safe until detonated using a blasting cap, which Alfred had invented a couple of years earlier. This was basically a plug of lesser explosive that could be ignited using a pyrotechnic fuse or an electric current. The blasting cap would then explode with enough force to detonate the Dynamite.

The new explosive, and subsequent developments, made Alfred a fortune. His business interests included investments in oil, his brothers Ludwig and Robert having become involved in developing the oil fields on the Caspian Sea, and he established a number of munitions factories. Dynamite, although it did have a military use in demolition, was mainly used for blasting in the mining and construction industries, but ballistite, one of Nobel's subsequent nitroglycerine-based inventions, was designed as a smokeless propellant to replace gunpowder in ammunition.

The surviving Nobel brothers, through their hard work and ingenuity, appeared to have made the tortuous journey from rags to riches, suffering their share of tragedy along the way. One day in 1888, however, Alfred was more than a little surprised to read his own obituary in a newspaper. The piece was titled 'The Merchant of Death is Dead' and was highly critical of him for having dedicated so much of his life to the armaments industry. Alfred, of course, was not dead. His brother, Ludvig, had died and the newspaper had got them mixed up. Appalled that future generations might be led to believe that he was some kind of monster, Alfred decided to establish a foundation that would reward those who bestowed the 'greatest benefit on mankind'. He altered his will to that effect, leaving the lion's share of his vast fortune to create the five Nobel Prizes – Physics, Chemistry, Medicine, Literature and Peace.

Without dynamite, invented by a man whose business interests had always revolved around arms and munitions, the world would be a far different place. There would be no Nobel Prizes and non-clumping kitty litter might never have come to pass!

NOT JUST
A SCOTTISH SWORD

(THE CLAYMORE)

Connor MacLeod will be a name familiar to anyone who knows the *Highlander* movies and subsequent spin-offs. MacLeod was a Scottish crofter, played by French actor Christopher Lambert, who meets an Egyptian traveller, played by Scottish actor Sean Connery, who informs him that he is immortal. The immortals are doomed to roam the Earth, unable to die or be killed unless they are decapitated, seeking each other out to do combat until only one of them remains. For Connor MacLeod, that combat begins with him using a claymore, a medieval Scottish broadsword, which he uses to lop off the heads of other immortals before they can do the same to him.

What the fictional MacLeod could really have done with was a different type of claymore – the Claymore mine, invented by another of his clan, the real-life explosives expert Norman MacLeod. The claymore couldn't be guaranteed to decapitate a rampaging immortal, but it would certainly have given them something to think about.

When the Allied troops first encountered Chinese tactics during the Korean War, they were shocked by the way that squads of Chinese attackers were directed against what were perceived to be weak points in Allied defences, pressing home their attack relentlessly, squad after squad, regardless of losses. Sometimes the Chinese commanders kept sending in their assault teams until the defenders simply ran out of ammunition. To counter this sort of determined attack, the Allies began looking at new kinds of weapons capable of defending set positions.

One such weapon was the Claymore mine. The Claymore was especially effective against groups of advancing troops because it used what is known as the Misnay-Schardin effect. Misnay and Schardin were two explosives experts during the Second World War who began looking at 'shaped charges', primarily to be used in anti-tank devices. The Misnay-Schardin effect is a term used to describe the way that an explosive blast travels directly away from the surface of the explosive charge. A flat sheet of explosive positioned vertically will send the blast outwards perpendicular to the sheet, both to the right and to the left. A round ball of explosive will send the blast outwards in all directions.

Defending a position using land mines buried in the ground, much of the blast tends to be lost upwards, only affecting an enemy above the mine (who has stood on it to detonate it) or very close by. MacLeod's idea for his Claymore was to use a curved-steel backing plate, with the inside of the curve facing towards the defender. A layer of explosive lined the outside of the plate and 700 ball bearings set in resin were placed in front of the explosive, the whole lot housed in a lightweight plastic case. When the explosive was detonated with an electric charge, the blast sent the ball bearings, their shape distorted into ragged missiles by the explosion, hurtling outwards in a fan shape. Because the Claymore sat on short legs above ground, the fan was over 6 feet (1.83 metres) high and 165 feet (50 metres) wide. The effect was lethal up to about 165 feet (50 metres). The mines could be placed along the route an enemy might be expected to take and camouflaged with foliage, carefully hidden control wires running back to the defensive position from where the mines could be detonated either individually or in sequence using an electric switch.

The Claymore was developed by MacLeod in cooperation with the United States Army and has been used to great effect in battle zones from Vietnam in the 1960s to Afghanistan in the 21st century. During its 60-year lifespan, the Claymore idea has been developed and refined. It has also been copied by the armies of numerous countries around the world including, of course, the Chinese.

It might not be combat the way that Connor MacLeod knew it, but the Claymore helps to ensure that hard-pressed defenders don't panic and don't lose their heads.

DRINKING
CLEAN WATER

(WATER PURIFICATION TABLETS)

Providing soldiers in the front line with adequate supplies has always been a logistical nightmare for any army. Sometimes it simply isn't possible to get food to personnel defending isolated positions. When the emergency field rations with which he has been issued run out, a soldier in the field can be relied upon to carry on functioning without food for a couple of days. He won't like it – who would? – and will complain bitterly, but he won't die. Any longer than two days, and he will have trouble concentrating or summoning up the energy to do his job, but he still won't die.

If a soldier has to go two days without proper drinking water, on the other hand, he will die. If he's operating in a hot climate, he may die a lot sooner than that, but even in mild conditions, after two days without water he will be too dehydrated to be of much use to anyone. A soldier needs to be able to find drinking water, but water that he comes across in a jungle pool or a shell hole on a battlefield might well be dangerous to drink. Bacteria in standing water can cause a range of problems from a simple upset stomach to a deadly disease such as typhoid or cholera.

Boiling water is a good way to kill off many of the bugs that might be lurking unseen in what looks like perfectly clear water, but in the field that would involve lighting a fire. Smoke from a daytime fire, and the flames from one at night, could easily give your position away to the enemy. In any case, lighting a fire might not be possible if there is no wood for fuel.

Looking for a way to give soldiers a fighting chance of using whatever water might be available to them in the field, Major Carl Rogers Darnell, Professor of Chemistry at the US Army Medical School, developed a way to purify water using liquid chlorine and during the First World War he designed a purification filter for use in the field. US Army Surgeon Major William Lyster came up with the 'Lyster Bag', a canvas bag holding 36 gallons of water to which chlorine was added to sterilise the water. Clearly, large bags and filters were for use by whole squads of soldiers, and individuals were issued with small phials of chlorine for personal use. When nothing else was available, household bleach was used.

Needless to say, none of these things made the water taste very good and none were particularly convenient to carry in battle conditions. In the latter stages of the First World War, halazone tablets, a chlorine compound in tablet form, became available and just over 20 years later, when the Second World War began, soldiers were regularly issued with halazone tablets to sterilize the water in their canteens. A bottle of tablets was light, adding little to the overall weight of a soldier's battle kit, and would last him for several days. During the 1940s, Harvard University cooperated with the US Army to produce another type of iodine-based water purification tablet.

Since then, various other types of 'clean water' tablet have been developed not only for military use but for the commercial market, to be used in all sorts of situations, from ramblers or campers exploring remote regions, to aid workers operating in areas where clean water is difficult to come by.

Clean water is something that most of us take for granted, but for many the simple water purification tablet intended to keep a soldier ready for battle has become a real life saver.

PLAYING WITH FIRE

(INCENDIARY BOMBS)

The ability to light a fire is something that helped early humans to set themselves apart from other animals and go on to create great civilisations and an industrial world, but for all the enormous benefits brought by fire, there have always been some fairly devastating drawbacks, too. It is horrifyingly effective as a weapon and being able to rain fire on your enemies gives you a battle-winning advantage.

Flaming arrows and pots containing potent mixtures of oil, sulphur and other materials are known to have been used in battle, against enemy settlements and in naval warfare, up to 3,000 years ago and any such concoction eventually came to be known as 'Greek fire', although Greek fire could also refer to an early type of flame thrower.

Over the centuries, the use of fire and incendiary devices became ever more sophisticated but it wasn't until the industrial age that an effective incendiary bomb was developed. The first incendiaries were dropped by German Zeppelin airships on the south coast of England in early 1915. The bombing was not accurate enough, and too few bombs were dropped to cause substantial damage but the raids certainly gave everyone living within range of the Zeppelins cause for alarm. The idea that your enemy could appear in the sky above your town and drop bombs on you was terrifying. Civilians living hundreds of miles from the front-line battlefields, across national borders and even across physical barriers such as the English Channel – long held as an unassailable bastion capable of keeping any nasty European war at arm's length – suddenly found that a modern war could reach out and touch them. That would have a devastating effect on morale that would be exploited to the full during the Second World War.

Although the bombing campaigns of the Second World War began in a limited way, they soon escalated to encompass the doctrine of 'total war', where any means of striking at your enemy become legitimate tactics. Incendiary bombs were developed that used aluminium, magnesium, titanium or other metal powder fuels to make 'thermite', a material that would burn at an extremely high temperature. The standard RAF incendiary was a 4 pound (1.8 kg) bomb that could burn for up to 10 minutes setting light to anything flammable nearby. Hundreds of these could be dropped from a Lancaster bomber along with huge, high-explosive 'cookie' bombs, weighing up to six tons, which were also known as 'blockbusters' because they could level an entire city block. The blast from the cookie was intended to blow the roofs off buildings, allowing the incendiary devices to set fire to the insides.

The RAF's tactic, which the Germans were never really able to match, even in the infamous London Blitz, was to concentrate their incendiaries so that the thousands of small fires would combine to burn out of control as one giant inferno. When Hamburg was hit by more than 700 Allied bombers during one raid in July 1943, it created the first firestorm. The fires heated the air above the city to temperatures in excess of 800° Celsius up to an altitude of 1,000 feet (300 metres). The hot air rose rapidly, to be replaced by a rush of cool air drawn in at ground level. The rush of air caused tornado-like winds to howl through narrow streets at up to 150 mph. This fed the fires with fresh oxygen and debris to help them burn even more fiercely. Anyone caught out in the street was also likely to be sucked into the flames. Even those sheltering from the maelstrom below ground in cellars were not safe as the fires used up all of the available oxygen. Many of those who evaded the flames died of suffocation. More than 40,000 people lost their lives in the firestorm.

On 10 March 1945, American B-29 Superfortress aircraft firebombed Tokyo. Numerous air raids had been launched against the city but this became the most devastating raid of all time. It is estimated that 100,000 people were killed – more than in either of the subsequent atomic bomb attacks – and 25 per cent of the city was destroyed.

In 1980 the United Nations Convention on Certain Conventional Weapons decreed that incendiary devices were not to be used against civilian targets or military targets located near civilian areas.

ONE HUNDRED MILLION AND STILL GOING STRONG

(KALASHNIKOV AK-47)

If there is one symbol of violent conflict that it instantly recognisable around the world, it is the stubby barrel and curved magazine of the Kalashnikov AK-47 assault rifle. The weapon is notorious for being used by rag-tag armies bent on bloody revolution, terrorists out to create havoc or pirates out for kidnap and plunder. It is a reputation that it doesn't really deserve as the AK is used as much by those defending legitimate regimes and protecting the rule of law as it is by the 'bad guys', but when around 100 million of the things are in circulation, some of them are bound to fall into the wrong hands.

Having his name attached to a creation that is so often associated with acts of pure evil is not what Mikhail Kalashnikov had in mind when he set out to become an engineer.

Kalashnikov's parents had been deported from their home in Altai Krai in 1930 during one of Stalin's great purges, when peasant farmers who were seen to have more land or livestock than their neighbours were denounced as 'kulaks' (tight-fisted enemies of the people), had their property taken away from them and were sent to live in poverty. As a result, Mikhail was a sickly child, almost died when he was six years old and remained of small physical stature as he grew into adulthood. Fascinated by all things mechanical, he left home in his early teens to hitch-hike hundreds of miles back to his original home where he became a mechanic, working on tractors. His aim in life, he maintained, had always been to invent ways of making a farmer's life easier.

In 1938, Mikhail was drafted into the Soviet Army, his size and skills making him an ideal candidate to squeeze into the confined spaces in a tank as a mechanic. He was rewarded for his efforts in devising improvements to tanks and other weapons, eventually becoming a tank commander, but was wounded in action at the Battle of Bryansk in 1941. It was while in hospital, listening to other wounded soldiers complaining about the poor quality and unreliability of their rifles, that Mikhail started thinking about designing better infantry weapons.

His first designs were not put into production, but they did earn him a place at the Red Army's weapons development unit where he submitted a number of weapons for consideration, elements of which eventually evolved into the AK-47 (Automatic Kalashnikov model 1947) which was adopted as the standard assault rifle for the Soviet Army in 1949.

The AK-47 was never intended to be a precision-made, sensitive, highly accurate firearm. It was designed to be used by Soviet soldiers wearing heavy gloves in battle conditions where snow, mud or grit would inevitably conspire to clog up the works and render the rifle unusable. That was the sort of thing the wounded soldiers Mikhail had heard in hospital were complaining about. The moving parts of the AK-47 were made so that they would keep on working when other weapons would have seized up. Most rifles have a tight, slick operation that stops the moving parts from rattling around and causing the weapon to buck violently when fired, spoiling the aim.

The AK-47 may have sacrificed a little in accuracy, but it made up for it by providing reliable, relentless firepower. It was rugged enough to stand up to the harshest of treatment and simple to maintain. Over the years, that is what has made it the weapon of choice for those who have little or no technical support in the field. It could be switched from semi-automatic, when it fired one round each time the trigger was pulled, to fully automatic when it would keep on firing until the magazine was empty. The standard, curved magazine held 30 rounds but the rifle would accept other magazines, including a round, drum magazine holding 100 rounds. The AK-47 could empty the drum in less than a minute.

Of the incredible 100 million AKs (the weapon has gone through several developments with different designations, although most people in the West refer to them all as the AK-47) that have been made, only 10 million were manufactured in the Soviet Union. The weapon was produced under licence in countries as far apart as Finland, Nigeria, Israel and Venezuela and copied to be produced illegally in countless workshops and factories around the world. More AKs have been produced since 1949 than all other assault rifle types put together.

In Russia, the latest version of Mikhail's invention, the AK-103, is still in production.

THE SURGICAL MIRACLE

(PLASTIC SURGERY)

In the 21st century, when 'boob jobs', facelifts, tummy tucks and all manner of cosmetic surgery are available to those vain enough to want to undergo the operation, and when reconstructive surgery for accident victims or patients with diseases that require corrective procedures is regarded as commonplace, plastic surgery is a term with which everyone is familiar.

'Plastic' in this context has nothing to do with the kind of plastic that is made in factories to produce everything from car dashboards to Lego bricks, although that could be said to have some relevance to boob jobs, but is actually to do with the meaning of the word in moulding or modelling. Despite the publicity that some celebrity recipients of cosmetic surgery generate, most plastic surgery undertaken nowadays is for the treatment of medical conditions or injuries, especially burns, where skin grafts to repair limbs or, most significantly, to reconstruct facial features are required.

While it seems like a thoroughly modern concept, corrective and reconstructive procedures can trace their origins back almost 3,000 years to pioneering surgeons in India but until the modern era any kind of surgery was an extremely risky business involving a great deal of pain with no guarantee of success – or even the patient's survival. The development of painkilling drugs, effective anaesthetics, antibiotics to combat infection and the sterilisation of surgical equipment all helped to make modern surgery viable. Aside from those injured in the many work-related accidents of the industrial age, the most abundant supply of patients requiring plastic surgery for the most appalling injuries inevitably came from the military.

At the beginning of the First World War, 32-year-old New Zealander Harold Gillies joined the British Royal Army Medical Corps. Having studied medicine at Cambridge University and worked at hospitals in London, the young surgeon helped to treat countless casualties in France and saw how small skin grafts could be used to patch up wounds. He returned to England where he persuaded the army that they should establish a special unit to deal with facial injuries. This eventually led to the opening of Queen Mary's Hospital at Sidcup in south-east London in 1917. It was here that Gillies operated on 27-year-old Walter Yeo. Yeo had suffered horrific injuries during the Battle of Jutland while serving aboard HMS *Warspite*. Much of the skin on his face was burned off and he lost his upper and lower eyelids. Gillies was able to use a new technique he had developed to take a flap of skin from Yeo's shoulders and use it as a mask across his cheeks, nose and eyes, creating a new nose and eyelids. The flap was not completely removed from the shoulders but remained attached by tubes of skin that ran from Yeo's face, round his neck to his back. These maintained the blood flow to the skin flap, keeping the graft alive in its new position until it became established. Warrant Officer Walter Yeo, therefore, became the first person to undergo modern plastic surgery. His disfigurement, although vastly improved, was always apparent but he survived, returning to serve in the Royal Navy in 1919 and living in Portsmouth until his death at the age of 70 in 1960.

Gillies and his team performed over 11,000 operations on First World War casualties at Queen Mary's Hospital, and over the next 20 years he toured the world lecturing on the techniques they had used and training new surgeons. He became Sir Harold Gillies in 1930 and during the Second World War he was a government advisor as well as continuing his surgical work with the army at Rooksdown House near Basingstoke. One of the many surgeons who trained under Gillies was his cousin, Archibald McIndoe, whose burns unit for treating RAF personnel at the Queen Victoria Hospital in East Grinstead became world famous. McIndoe developed and expanded his cousin's techniques while inventing many of his own, his patients forming what they called 'The Guinea Pig Club' such was the experimental nature of much of his plastic surgery. The club was for severely burned airmen who had undergone ten or more surgical procedures. By the middle of 1941 the club had just 39 members, including the medical staff who operated on the men. Four years later there were 649.

The groundbreaking work of Sir Harold Gillies and Sir Archibald McIndoe (knighted in 1947) in treating military personnel undoubtedly paved the way for the plastic surgery that is routinely undertaken today, every hour of the day, all over the world.

YOU'RE IN TROUBLE WHEN THE NAGGING STOPS ...

(SATELLITE NAVIGATION)

That annoying voice that you hear in the car as you're driving along (no, not the one from the back saying 'Are we there yet?') telling you when to turn right or 'Enter roundabout and take the second exit' has become as much a part of motoring as the mortgage you need to fill up your tank or the youth waiting at the traffic lights to relieve you of your loose change in return for smearing your windscreen with a damp rag. The voice of the satellite navigation system, and the advice it imparts, have become both a boon and a burden.

Before we became slaves to the sat-nav, we could enjoy the pleasure of poring over a map before setting out on a long journey, making copious notes about where to turn off the motorway and how to avoid the mind-numbing limbo of a city-centre traffic jam. Now, sat-nav will do it all for you, finding you the fastest route, suggesting alternatives if you take a detour and even warning you where to look out for speed cameras into the bargain. None of that would be possible without military technology.

The idea of satellite navigation, or at least finding your way in the dark to some place you've never been before without using a map or asking for directions, really began during the Second World War.

Navigating at night, or even during the daytime for that matter, to try to find a specific factory on a specific river hundreds of miles away, drop your bombs on it and find your way home again, was no mean feat when you were trying to work from maps or charts, judge the strength of the wind blowing you off course, judge your position by picking out landmarks four miles below you, estimate your speed and time your turns using a clock and a compass. It's little wonder that bombers often ended up dropping their payloads miles from where they were meant to be, fully believing that they were bang on target.

One of the things that changed that was the use of radio beams to direct bombers. The Germans used a system where they broadcast radio signals from transmitters in Scandinavia and Germany. The bombers were able to listen to the tone of the signal as they flew out along the radio beam. If they started to lose the signal, they knew they had to adjust their course until they were receiving it full strength. As they approached their target, they would start to pick up the second signal. This reached maximum strength when the two radio beams crossed above the target and they could then release their bombs.

The RAF used a similar system that measured a time delay between two radio signals to give the navigator a 'fix' on his position. This is actually far more akin to satellite navigation, the biggest difference being that the signals used by a sat-nav system to calculate its position come not from radio masts on hilltops but from outer space.

The first satellite navigation system was called Transit, and was developed by the United States military in the 1960s. Its purpose was to help the US Navy's nuclear missile submarines to figure out exactly where they were. The satellites transmitted set signals and a Transit receiver could calculate where it was, based on several satellite signals and their known orbits. Transit entered service in 1964 and was still in use up to 1996, by which time the more sophisticated GPS was in operation, using a network of two dozen military satellites.

GPS (Global Positioning System) satellites transmit information about their own position, meaning that the receiver doesn't need to have prior knowledge of the satellite's orbit schedule, and the calculations about the receiver's position can be made far more quickly and accurately using far smaller receiving units. Today's mobile phones are equipped with GPS systems that would have amazed the scientists who set up Transit half a century ago!

The civilian applications of military satellite communications and navigation technology are manifold but the most obvious for many of us is the sat-nav system nagging away at us in the car. However much you rely on your sat-nav, it might be wise always to keep a map handy as a back-up. In certain parts of the world, when major military operations are underway, the satellites' military masters can switch their signals to a special frequency to give them sole use of the system, meaning that the nagging voice coming from your dashboard will suddenly fall silent ...

DIY FOR
OUTER SPACE

(CORDLESS POWER TOOLS)

When two very down-to-earth young American businessmen established a machine shop in Baltimore, Maryland, in 1910, they had plans that would make their fortune but S. Duncan Black and Alonzo G. Decker could never have imagined how their Black and Decker Manufacturing Company would become a household name all over the world and that Black and Decker tools would even be used on the Moon.

Black and Decker produced tools for industry, such as a capping machine for a bottling plant, adding machines, a dipping machine for making sweets and machinery for the United States Mint. It wasn't until 1917 that they patented their hand power drill with a pistol grip and trigger switch. By the end of the 1920s, the company had grown enormously, establishing factories in Canada, the United Kingdom and Australia.

Alonzo Decker Jnr, who had worked at the company co-founded by his father since he was 14 years old (except when, to show how fair the company was, he was one of the first to be laid off during cutbacks in 1931) noticed a strange sales pattern when supplying power tools to defence contractors during the Second World War. Repeat orders were coming in at a rate that suggested the tools might be faulty. Far from it. The defence contractors' employees were ordering power tools for themselves to take home to do jobs around the house. If people wanted them at home, Decker reasoned, then there was a commercial market for Black and Decker products on the domestic front as well as in industry. Supplying power tools for the home was instrumental in creating the post-war DIY boom. Affordable power tools for the home meant that there really was no excuse not to fix those new shelves or screw those kitchen cabinets to the wall. The reluctant home handyman has Alonzo to blame ...

In 1961, Black and Decker introduced the first cordless power drill, which was a boon to tradesmen working on building sites where there was no electricity, but the weight of the batteries and the fact that they drained so quickly made cordless tools of limited value to the general market. NASA, however, saw great potential in cordless tools, but they needed tools for use in outer space to be lightweight, robust and capable of operation without constant recharging.

Black and Decker worked with the Martin Marietta Corporation, funded by NASA, to develop their cordless tools, testing them underwater and in the holds of diving transport planes to simulate zero-gravity. For the Gemini programme, NASA needed a power wrench that would undo bolts in outer space, turning the bolt heads without spinning the astronaut in the opposite direction, and for the Apollo Moon missions, they needed a drill that would extract rock samples from the lunar surface. There is, after all, nowhere to plug in your drill on the Moon.

Work on creating small, lightweight, cordless power tools for NASA led to the development of other hand-held cordless devices such as a miniature vacuum cleaner called the Dustbuster as well as precision medical instruments and surgical tools. NASA may not have devised the very first cordless drill, but the technology developed for them brought about the cordless revolution, resulting in devices that are now used on a daily basis both in industry and in the home.

SPECIAL FORCES FLASH-BANG

(THE STUN GRENADE)

The stun grenade, also known as a 'flash-bang', is a weapon devised by Britain's elite special forces regiment, the Special Air Service (SAS).

When they began training in what they termed Counter Revolutionary Warfare (CRW) tactics in the 1960s, subsequently establishing their anti-terrorist team, the SAS realised that they needed to be able to clear a room in a building using a new approach. In a typical combat situation when the enemy had ensconced themselves behind four walls, the way to flush them out was to lob in a couple of hand grenades then kick the door in as the dust settled and spray the room with machine-gun fire. The likelihood of anyone inside the room surviving was rather slim. As an anti-terrorist team, however, they would not be operating under battlefield conditions and were more likely to be facing opponents holed up in a room or building with civilian hostages. The objective would be to rescue the hostages, not simply eliminate everyone in the room.

Part of the solution to the problem was the stun grenade. A normal grenade is designed to spray fragments of metal from its casing in all directions – hence it is known as a fragmentation grenade – with this shrapnel causing horrendous injuries to anyone in close range, especially in an enclosed environment. But even those inside a room where a fragmentation grenade had been used were left somewhat dazed by the blast and the noise, often disorientated for long enough to allow attackers to storm in and shoot them before they could react. So what if there was no deadly shrapnel, just a deafening bang and a blinding flash? The stun grenade was designed to do just that.

Keeping things as simple and familiar as possible, the stun grenade was operated by pulling a safety pin and releasing a striker lever that activated the fuse which in turn set off the detonator. A mixture of magnesium or aluminium powder along with ammonium or potassium compounds then exploded to provide a bright flash and a loud bang.

Unlike a fragmentation grenade, the housing of a stun grenade is not designed to disintegrate into flesh-ripping shrapnel. As few metal parts as possible are used and in some cases the grenade is built to remain intact, with holes drilled in the casing to let out the light and sound.

The light produced by the SAS's original flash-bang was equivalent to around 300,000 candlepower – bright enough to blind anyone for a few seconds. The bang was around 160 decibels, roughly equivalent to a shotgun blast right next to your ear. Together the flash and bang were enough to blind, deafen and disorientate someone for between three and five seconds – long enough for the SAS to gain entry, identify friend from foe and take appropriate action.

The stun grenade was first used operationally when two SAS soldiers accompanied a German special forces unit storming a hijacked Lufthansa airliner on the runway at Mogadishu in 1977. They were also used by the SAS to great effect in their assault on the Iranian Embassy in London during the famous siege there in 1980.

Since they were first used by the SAS, various kinds of flash-bangs have been developed and are now in use by police forces and military units all over the world.

A NUCLEAR SOLUTION

(THE ATOMIC BOMB)

Never has there been a weapon that has inspired as much fear as the atomic bomb, a hideously remarkable achievement considering that nuclear weapons have been used operationally just twice.

At the turn of the 20th century, scientists were beginning to understand far more about the structure of atoms and how to manipulate them. In 1898 Marie Curie had discovered a substance that she called radium which emitted radioactivity. Her subsequent experiments with radioactivity and X-rays, when their effects on the human body were completely unknown, led her to suffer from a variety of ailments caused by radiation exposure and caused her death in 1934. Her research papers and even her cook book are now considered too radioactive to be handled unless by someone wearing protective clothing.

The discoveries made by Curie and others led a number of researchers to conclude that there were vast amounts of energy stored in materials such as uranium that could be released under the right circumstances. Scientists in the United States and Europe developed techniques that allowed them understand how atoms are structured and, from there, how to split certain atomic particles, releasing energy. Leó Szilárd, an American physicist of Hungarian extraction working in London, saw the destructive potential of a chain reaction and filed a British patent for his nuclear chain reaction in 1934. Within two years, he had assigned his patent to the British Admiralty, realising that it had the potential to create a weapon of enormous destructive power. As an Admiralty patent, the details could be kept top secret.

Szilárd, however, was not the only one to come to this conclusion. German researchers Otto Hahn and Fritz Strassmann reported in 1938 that they had bombarded uranium with sub-atomic particles called neutrons and then detected the presence of a new element, barium. This process of transforming one element into another by breaking it down is known as nuclear fission, and it involves the release of vast amounts of energy. In 1939, when the Second World War began, the Allies had grave concerns that the Germans would be able to develop a superbomb that utilised the effects of nuclear fission. Despite some early rivalry, Britain and the United States eventually agreed to work together and British and Canadian scientists, who had been working on a project that was codenamed 'Tube Alloys', joined forces with the American 'Manhattan Project' to create a highly secret research organisation that ultimately involved 130,000 people working at sites throughout the United States as well as in Canada and the United Kingdom. The majority of resources were supplied by the United States to ensure that the Allies won the race to create the atomic bomb.

Julius Robert Oppenheimer, Professor of Physics at the University of California, Berkeley, was chosen to establish a laboratory at Los Alamos in New Mexico where the first atomic bomb was designed. The bomb was tested at a US Army bombing and gunnery range in New Mexico, now the White Sands Missile Range, on 16 July 1945, two months after

Germany had surrendered. The war in the Far East, however, was still raging. Japan had refused to surrender and the casualty rate involved in 'island hopping' across the Pacific to bring the Americans within striking distance of the Japanese homeland was so high that the idea of actually invading Japan was a daunting prospect. The American military planners estimated they would suffer 50,000 casualties on the first day of an invasion, with the Japanese taking many times that number. A full-scale invasion may have resulted in a million deaths. The decision was taken to use two atomic bombs.

The first was dropped by a B-29 aircraft on the city of Hiroshima on 6 August 1945. The bomb destroyed almost five square miles of the city, killing up to 80,000 people immediately and injuring a further 70,000. Many would later die from burns or the effects of radiation. The second bomb, again dropped from a B-29, was used three days later against the city of Nagasaki where 35,000 people died and 60,000 were injured. Faced with the prospect of more of these devastating weapons being used against them, the Japanese began negotiating surrender terms on 10 August.

In the post-war era, as East faced West in a clash of ideology, the threat that nuclear weapons could be used was one of the factors that prevented the Cold War turning very hot.

REACHING BEYOND

(THE BALLISTIC MISSILE)

The Second World War brought technology that previously only existed for most people in the realms of science fiction and fantasy stories very much into the real world, with one particular example being the first man-made object actually to leave our world – the ballistic missile.

During the Cold War, the political superpowers of the Eastern Bloc and the Western Allies stockpiled missiles capable of hitting targets thousands of miles away, with nuclear warheads that would have devastated the entire world. Everyone knew that a full-scale nuclear duel between East and West had the potential to make our entire planet uninhabitable and the stalemate of what was appropriately referred to as M.A.D. (Mutually Assured Destruction) was maintained by the opposing factors keeping a very close eye on each other's nuclear arsenals to ensure the power balance was not upset in any way. While bombers and even nuclear artillery pieces formed part of those arsenals, the premier delivery system for a nuclear device was the ballistic missile.

'Ballistic' is simply a word that describes something that flies through the air unaided, obeying the general laws of physics. A ballistic missile is a projectile that is provided with an initial boost but then follows a flight path largely dictated by the forces of nature. You might think of the earliest ballistic missiles as being stones or spears. Whoever was throwing them sent them on their way, but after that they were on their own. The Chinese were using 'fire arrows', developed from their fireworks and using gunpowder as the propellant, more than 1,000 years ago and various types of rocket were used as artillery over the years with varying degrees of success. It wasn't until 1944, however, that the modern ballistic missile was used for the first time.

Hitler's *Vergeltungswaffen*, 'revenge weapons', were what he hoped would persuade the Allies to sue for peace during the latter stages of the Second World War. The V-1 was a kind of cruise missile powered by a pulse-jet engine which went ballistic when the engine cut out over the target and it dropped to earth, delivering its one-ton explosive payload. The V-2 was a different beast altogether. Developed by rocket scientist Wernher von Braun and his team, the V-2 used a liquid-fuelled rocket engine that blasted it up into the air to an altitude of more than 60 miles, taking it out of Earth's atmosphere. The missile's engine would cut out after a minute or so, depending on the range to the target, and it would follow a curved flight path, dropping down into the atmosphere with its speed rising to four times that of sound as it fell through the air. The arrival of a V-2 on target was heralded by a supersonic double thunder-clap before it hit the ground. Its high-explosive warhead created a huge crater, throwing around 3,000 tons of rubble into the air.

Unlike the V-1, which could be intercepted by Allied aircraft and shot down, there was no defence against the V-2. The first one was used against Paris in September 1944, with 1400 subsequently launched against London and 1600 falling on Antwerp, amongst other targets.

Just eight months after that first operational launch against Paris, Wernher von Braun and his team surrendered to the Americans, fearing capture by the Russians, as the Allies advanced into Germany. Von Braun's knowledge and expertise were invaluable in developing the ICBMs (Inter Continental Ballistic Missiles) of the Cold War and in creating the rockets that would ultimately take astronauts to the Moon.

BRINGING OUT THE BIG GUNS

(THE BATTLESHIP)

If a government wanted to flex its muscles and flaunt its economic, industrial and military might, perhaps to show that diplomats engaged in meaningful talks with foreign rivals should be taken seriously, there were, at one time, few bigger sticks to brandish than a battleship. When a giant warship sailed into port bristling with guns that guaranteed enough firepower to destroy a city, everyone sat up and took notice.

The word 'battleship' was first used more than 200 years ago as a short form of the term 'line of battle ship'. The 'line of battle' referred to the naval tactic from the days of sail when admirals went into battle with their most powerful ships in a long line, cruising past the enemy formation side-on at close range to allow them to bring the maximum number of cannon to bear, the guns firing through ports in the side of the ship. A line of battle ship was also known as a 'ship of the line'. The only ship of the line to survive today as a commissioned warship is HMS *Victory*, Nelson's flagship at the Battle of Trafalgar, still serving in the Royal Navy, albeit from her berth in drydock at Portsmouth, almost 250 years after she was launched.

The line of battle tactic was to become obsolete by the middle of the 19th century when warships began to use steam power.

Steam ships did not have to rely on the wind and could sail in any direction they chose, forcing a rapid re-think of naval battle strategy, although the name 'battleship' remained in use, referring to the largest, most heavily armed ships in the fleet.

The coming of the industrial age led to a revolution in shipbuilding just as it did in every other manufacturing industry and the old, wooden, wind-powered warships gave way to 'ironclad' wooden ships with armour plating, soon to be followed by ships built of iron and steel armed not with simple cannon but with massive naval guns that fired explosive shells thousands of yards. In 1876 the French battleship *Redoutable* became the first steam-powered warship made entirely of iron and steel.

As the dawn of the 20th century approached, the industrialised world was engaged in something of a naval arms race that moved into top gear with the launch of the Royal Navy's HMS *Dreadnought* in 1906. *Dreadnought*'s steam turbine engines gave her a top speed of 21 knots, the fastest battleship in the world, and her main armament consisted of ten 12-inch guns mounted in five of the latest rotating gun turrets. *Dreadnought* was the first of the modern era of battleships and, conversely, the modern era meant the end of the line for the battleship.

While there were some clashes between battleships during the First World War, notably at the Battle of Jutland, the huge ships seldom went head to head. It also became clear that battleships were vulnerable to attack by smaller, more nimble ships firing torpedoes and by stealthy submarines. The enormously expensive battleships could even be sunk by a relatively cheap exploding mine. Bizarrely, battleships needed to be protected by their own flotillas of destroyers and, when aircraft began to be used in earnest, they faced yet another hazard every time they put to sea.

Yet the battleship still had its uses. During the Second World War, the German battleships *Bismarck* and *Tirpitz* were seen as such a threat to the Atlantic convoys keeping Britain supplied with food and munitions that massive efforts were made to hunt them down and sink them. The aircraft, ships and personnel that were lost combating the threat from the two German battleships were a huge loss to Britain. The battleship's role as a mobile artillery force, however, was invaluable. Six battleships formed the backbone of the fleet that pounded the German defences in Normandy during the D-Day invasion and the mighty guns of the US Navy's battleships paved the way for seaborne assaults in the Pacific as the Americans flitted from island to island, closing in on Japan.

Most battleships were decommissioned after the Second World War but a few remained in service, or were brought back into service when their big guns were required for shore bombardment. Battleships were used to support landings in this way during the Korean War (1950-53) and in Vietnam. The USS *Missouri*, which served during the Second World War – the Japanese actually signing the document of surrender on her deck in Tokyo Bay in 1945 – saw action in Korea and shelled Iraqi positions during the Gulf War in 1991, using her trusty 16-inch guns as well as launching Tomahawk cruise missiles that would have been the stuff of science fiction when she first put to sea almost 50 years earlier!

Battleships have now been phased out of service with modern navies, replaced by a variety of more specialized weaponry, including aircraft carriers. In the modern world, if a government needs to lend its diplomats a little support, there's no need to send a battleship into port: a fly-past of jets from an aircraft carrier miles offshore will do the trick.

YOUR TEMPERATURE IN THE STARS

(THE EAR THERMOMETER)

Feeling a bit peaky? Think you might be about to come down with something and deserve a day off work? Best take your temperature to check. That, of course, always used to mean finding that delicate glass thermometer tucked away in the bathroom cupboard and sticking it under your tongue for a couple of minutes – not such a problem for most adults, but for an anxious parent trying to take a small child's temperature, it could be a nightmare.

Even if you did manage to keep the thing in place long enough to get a reading, you would then struggle to see just what it was trying to tell you, unless you could get it to catch the light in just the right way ...

For medical staff, the problem could be even worse. Even when a nurse threatened to use a rectal thermometer, it wasn't always enough to ensure complete cooperation from a belligerent patient. Spread that one time-wasting temperature-taking scenario across an entire country like the United States, where millions of temperatures are taken in hospitals every day, and a time-management accountant would be able to tell you how many wasted man-hours that entails, and the precise cost to the economy.

Then, in the early 1990s, the whole procedure for taking a temperature became far easier when the ear thermometer was introduced. An ear thermometer produces an accurate temperature within a couple of seconds, so instead of trying to hold an agitated child still long enough to see the mercury in an old-fashioned thermometer creeping up the glass tube, all you had to do was distract the patient for a moment while you rested the device in his or her ear. A digital display then gave you a precise reading of the patient's temperature.

The ear thermometer works by measuring the energy given off by your ear drum. As this energy – heat – is radiating from a source inside the body, the temperature reading is highly accurate. The device uses an infrared sensor to measure the energy emanating from your ear drum, which might not at first seem to have much to do with military technology – but the thermometer was actually developed by a company working with NASA's Jet Propulsion Lab, using technology that NASA had employed to determine the temperature of stars.

From advanced technology designed to help explore the universe and tell us more about celestial bodies millions of light years away, to something you can stick in your ear to tell you whether you're poorly enough for a day off work, NASA can be relied upon to come up with the goods!

CLEARING
THE FOG OF WAR

(THE CANNON)

Nothing has ever made quite such an impact on the battlefield as the cannon. The lethal effect of artillery fire was never more apparent than during the First World War, when the big guns of both sides were responsible for more combat fatalities than any other form of weapon.

The term 'fog of war', describing the confusion and uncertainty that erupts during a conflict and confounds military commanders, was first used by Prussian military analyst Carl von Clausewitz in the 19th century. While the phrase actually refers to the barrage of information and misinformation through which a commander must plough in order to direct his forces, the cannon, one of the commander's most valuable assets, contributed to the confusion in a very physical way. At a time when battles were conducted by generals who sought high ground from where they could view the battlefield in order to issue orders about when their cavalry should charge or when an infantry formation should advance, the thick, drifting smoke from the gunpowder used in cannon and muskets often created a very real 'fog', preventing anyone from seeing what was happening. This situation was mitigated somewhat by Alfred

Nobel's invention of 'ballistite' and the British 'cordite', derived from ballistite, both smokeless propellants that replaced the gunpowder, or black powder, that had powered cannon for a thousand years.

Since black powder was invented by the Chinese, it's hardly surprising that the earliest types of cannon were developed there as well. The first cannon were known as fire lances and were simple bamboo tubes attached to the ends of spears. The tubes were filled with gunpowder which produced a kind of flame thrower effect when the powder was ignited – a bit like pointing the rear end of a firework (also a Chinese invention, of course) at someone. It was soon realised that small stones or shrapnel packed in with the gunpowder would come hurtling out along with the flames, turning the fire lance into a cannon. Stuffing explosives into bamboo, of course, could be every bit as dangerous for the user as it was for the target if the whole thing unexpectedly went off with a bang. The answer was to create stronger tubes made of metal.

The first metal cannon were made of either iron or bronze. A charge of gunpowder was shoved down the barrel, followed by an iron or lead ball, some shrapnel or stones. The gunpowder was then ignited through a tiny hole at the rear, closed end of the cannon. Although initially these cannon were small enough to be carried in your hand, larger versions able to fire a bigger projectile over a longer range quickly followed. More power, however, made it more likely that the exploding gunpowder charge would blow your cannon to bits rather than simply firing the cannon ball, so as the cannon became bigger, the castings also became thicker, with the thickest part at the base, or breech. By the time of the Ming Dynasty (1368-1644) the Chinese had more than 3,000 cannon defending their Great Wall against the Mongols and the ships of their naval fleet were also armed with cannon.

Use of cannon had spread from the Far East through the Middle East to Europe by the 12th century but it wasn't until the Battle of Crecy in 1346 that an English army took artillery into battle. Widespread use of cannon by the northern Europeans is significant because of the way that they were to develop tactics for its use and, more importantly, because of the technical innovations that they would introduce. The early Chinese cannon were cast as hollow tubes, sealed at one end. The big problem with this was in pouring molten metal into a mould accurately enough to get the hollow part, the bore of the cannon, straight, with no deformities. As industrial Europe applied its technology to manufacturing arms, they began casting cannon as solid bars of metal, then drilling out the barrel, a far more accurate process.

In attempting to eliminate that smoky 'fog of battle' in the 19th century by replacing black powder, the Europeans also made other technical advances possible. The black powder residue clung to the inside of the barrel after a while, meaning that it had to be cleaned in order that you could carry on firing at the enemy. Cordite caused far less fouling, meaning not only that the barrels remained gunge-free, but also that they could be 'rifled'. It was already known that cutting grooves into the inside of the barrel would cause the projectile to spin, giving it gyroscopic stability in flight that made it far more accurate. When the British took their new, improved cannon back to China during the Opium Wars of the mid-19th century, the application of evolving western technology to the ancient eastern weapon was integral to the Chinese defeat.

Development of the cannon continued throughout the Industrial Revolution. Iron and bronze were replaced by steel. Accurate engineering allowed the introduction of breech-loading guns. Ramming a powder charge down a hot barrel had always been a nerve-wracking business for the gunners as there was always a chance that a smouldering scrap of wadding might blow the charge while the gunner was still stuffing it in with his ramrod. Breech loading was faster and safer for the gunners, especially since new forms of ammunition had been developed. Gone were the old cannon balls. Rifling had encouraged the use of shells, cone-shaped missiles packed with explosives that flew straight and true. They were often combined with a cartridge case containing the propellant so that both could be loaded together, greatly improving the cannon's rate of fire. If that new ammunition sounds just like the sort of thing you see soldiers loading into their rifles, then you might be wondering what is the difference between a rifle, or a pistol, and a cannon. Essentially, there is none. They do exactly the same job, but a firearm with a calibre (the diameter of the bore and, therefore, the size of the ammunition) greater than a heavy machine gun – 0.5 inches (about 12.5mm) – may be referred to as a cannon.

Gunners at the Battle of Waterloo in 1815, who couldn't expect to hit anything if it was more than a mile away, could scarcely have imagined that a century later, during the First World War, the Royal Navy's 18-inch guns would have a range of more than 23 miles and a huge German gun, so big that it could only be moved around on railway tracks, would bombard Paris from 76 miles away.

Cannon, more usually referred to nowadays simply as 'guns', are still being used. The guns of tanks and artillery pieces bombard enemy positions ahead of infantry attacks; quick-firing, multi-barrelled cannon are used on aircraft; and naval vessels, having replaced their large-calibre guns with guided missiles, still mount deck guns capable of delivering a shell to a target 20 miles distant. Modern radar range finding and target acquisition systems mean that the 21st-century cannon is more accurate than ever before and new naval cannon under development will fire rocket-propelled, satellite-guided warheads almost 120 miles, at a fraction of the cost of a cruise missile.

SLICE OF APPLE, ANYONE?

(THE CROSSBOW)

Most people know of the 14th-century legend of William Tell, the Swiss mountain man who, along with his young son, was sentenced to death for defying an aristocratic tyrant named Gessler unless he could shoot an apple balanced on his son's head. He did so, using his trusty crossbow, and inspired a rebellion that ultimately created modern Switzerland. Had Tell been using any other kind of bow, the result might well have been very different.

A traditional crossbow works on the same principle as any other kind of bow-and-arrow weapon. You draw a string that bends the arms of a bow inwards under extreme tension. The 'arrow' projectile is placed in front of the string and, when the string is released, the arms of the bow spring back into their original shape with great force, snapping the string forward and hurling the projectile towards the target.

The big difference between a crossbow and a normal weapon is that the crossbow is generally smaller, but delivers more power to the projectile. A normal bow relies on the strength of the bowman pulling on the string to bend the arms of the bow. The arms of the crossbow are shorter and need much more than a normal archer's single-arm pull to bend them. Some crossbows could be bent by bracing them against the ground and pulling with both hands; some used a hook on the bowman's belt to pull on the string while he had his feet on the weapon and straightened his legs; some used a lever mechanism or a ratchet device where turning a handle drew back the bow.

The bow string is then held in the 'cocked' position by a trigger mechanism. The trigger is mounted on a stock, a sturdy structure strong enough to support the bow when it is drawn. The projectile, called a bolt, sits in a slot on the stock waiting for the bow to be released and send it flying forwards. The bolt is generally shorter and thicker than a normal arrow and doesn't have feathered flights. The bowman can hold the butt of the stock to his shoulder and aim down the length of the stock, and hence down the length of the bolt.

Crossbows have been around in various shapes and forms, from small, pistol-type affairs to giant bows the size of field guns, for more than 2,500 years. Some of the famous warrior statues of the Terracotta Army in the tomb of Qin Shi Huang were armed with crossbows. Given that they are still used today by sports bowmen and still also have military applications, what is it that has made the crossbow such a success?

The whole point (pardon the pun) about a crossbow is that the bowman can take his time choosing a target. When an archer draws back a normal bow, he has to use his own strength to hold the tension while he takes aim. Unless he acquires a target quickly, his arms will start to shake and he won't be able to aim properly at all. A bowman with a crossbow has no such problem. He's not using his own muscle power to keep the bow bent, so he can afford to spend a little time making sure his aim is true. The crossbow doesn't have the same rate of fire as a normal bow, but neither do the bowman's arms start to run out of strength after repeated firing the way an archer's might.

What this meant was that, when a nobleman needed to raise a peasant army, he could equip some of them with crossbows and they could learn how to use them within a few days. For an archer to become proficient took strength, stamina and a great deal of practice, all of which required a lengthy period of training. Good archers tended to be sons of archers, who had been trained for years by their fathers. Crossbowmen could be committed to the battlefield almost straight away and in medieval Europe this certainly gave mounted knights something to think about because a well-aimed crossbow bolt was easily capable of penetrating their armour. A knight with sword or lance was no match for a peasant with a crossbow. England's King Richard I, 'The Lion Heart', was killed by a crossbow bolt in France in 1199.

William Tell would have been in real trouble if he'd succumbed to a dose of the shakes whilst trying to aim a normal bow at the apple on his son's head, and his son might have ended up with more than apple juice running down his face. Although we like to think that he trusted to his skill and his crossbow, legend has it that he had a second bolt at the ready, having decided to shoot the despicable Gessler if things went wrong. Ultimately, at a later date, he did use the bolt to assassinate Gessler and help to bring about a revolution.

POMEGRANATES AND OTHER FRUIT

(THE HAND GRENADE)

The pomegranate is an odd fruit, stuffed with large seeds that can make it challenging to eat, despite the fact that it is widely recognised as being good for your health. The pomegranate may be good for you, but it shares its name with something that is most definitely not – the hand grenade.

Pomegrenate is the old French word for the fruit but it is thought to be from where the word 'grenade' derives because its round shape is like the hollow balls that formed early grenades and the way the seeds are packed into the fruit is reminiscent of the way that grains of gunpowder were packed into grenades.

The hand grenade was not, however, invented by the French. Like most other things that go 'Bang!' the grenade was first invented by the Chinese as yet another way to make use of their gunpowder. A thousand years ago, Chinese soldiers were hurling ceramic or metal containers packed with 'divine powder' at their enemies. A burning fuse served to detonate the grenade, hopefully far enough away from the thrower for him to escape the blast and flying shrapnel. The ancient Greeks had previously experimented with a very effective incendiary substance known as 'Greek fire', probably a mixture of pine resin, sulphur and other elements known to burn well, which they poured into jars to hurl at enemies, but this would appear to be more like a Molotov cocktail weapon than the explosive-packed hand grenade.

When the idea of the hand grenade reached Europe some time around the 15th century, they had developed into small iron balls about the size of a tennis ball. A fuse had to be lit and the grenade thrown with impeccable timing. If you threw the thing too soon, the fuse would not have burned down and the enemy would have time to throw it back at you. If you hung on to it too long it was liable to go off in your hand.

It seemed like a good idea to give the job of throwing a grenade to someone who had the sense to know when to light the fuse, the nerve to stand up in front of the enemy and the physical strength to be able to throw the grenade far enough. Soon every regiment had its grenade throwers, generally tall, strong, athletic men, who were known as 'grenadiers', and by the 17th century entire regiments of grenadiers had been formed. These were considered to be elite soldiers – the bravest and strongest – and units such as the British Grenadier Guards, the most senior infantry regiment of the British Army, still retain a burning grenade as their cap badge.

By the time the First World War came round in 1914, a more modern style of hand grenade had been adopted, generally using a trigger system that activated a chemical fuse allowing for a five- or seven-second delay before the fuse set off the explosive. Many types had a stick handle that helped the thrower to hurl it a bit further. The British produced the Mills Bomb with a ring safety pin that had to be pulled before releasing a spring-loaded trigger. Because of its oval shape and its deeply notched surface (erroneously thought to aid fragmentation, producing more shrapnel), the Mills Bomb was nicknamed 'the pineapple' – another deceptively innocuous, fruity name for the deadly hand grenade. Around 75 million of them were produced during the First World War.

Although lethal up to a certain range when used in the open on the battlefield, hand grenades are devastating weapons when used in an enclosed space, such as a room in a house, the blast and shrapnel then having the greatest effect. Their use has had a huge influence on infantry combat tactics, but their effectiveness in urban situations means that hand grenades of various types, some producing smoke or debilitating gas rather than a lethal explosion, are valuable assets for police units operating in trouble spots all over the world.

POPEYE'S
FAVOURITE SUV

(THE JEEP)

There are some products that we all know and love so much that we simply adopt their name for everything that comes close to doing the same job. Brand names that have grown to become generic terms in this way include Sellotape for any kind of sticky tape and Hoover for any kind of vacuum cleaner. These brands have even grown into verbs with people talking about 'sellotaping an envelope shut' or 'hoovering a room.' Then there are brand names like Jacuzzi and Escalator that have become such commonly used terms that no one really recognises them as brands any more.

One brand name that falls into that category is Jeep. People refer to off-road vehicles or SUVs (Sports Utility Vehicles) as Jeeps, whether the car in question is actually made by Suzuki, Toyota, Daihatsu or even Land Rover. The Land Rover, in fact, is the world's second oldest four-wheel drive SUV, the oldest being the Jeep which was, of course, invented for the military.

The Jeep was a vehicle that nobody believed could ever be built. When the United States Army decided to update its transport fleet in 1940, having concluded that it was only a matter of time until it was drawn into the war that was already well underway in Europe, they asked 135 American motor manufacturers to design them a new truck.

The requirements for the small truck were very specific. It had to have a wheelbase of 75 inches, later extended to 80 inches, with the axles no more than 47 inches wide. Onto that basic platform had to fit an engine and transmission system that would drive all four wheels and a body that would carry at least three fully armed soldiers and a .30 inch calibre machine gun. Without passengers or equipment aboard, the little truck was to weigh no more than 1,300 pounds (590 kg) so that it could easily be hauled out of muddy ditches; it also had to be able to carry its load up a 45° slope and splosh through water 18 inches deep.

Because they knew they would have to be ready to go into battle in the not-too-distant future, the army demanded to see a prototype in less than two months with a fleet of vehicles ready for testing less than a month after that. Only three companies rose to the challenge – Ford, Willys Overland and American Bantam.

Bantam were known for having produced the British Austin Seven under licence but had fallen on hard times. Even so, they had their 'Blitz Buggy' ready for testing before Ford or Willys. Both of the other companies had representatives present when the army tried out the Blitz Buggy and those representatives were sent away with blueprints for the Bantam design to make sure that they came up with something at least as good. Willys eventually produced the 'Quad' and Ford came up with the 'Pigmy', both of which were put through their paces alongside an improved Bantam. Unable to make a definite decision, the army ordered 1,500 of each truck to be delivered to the units that would be going to war in them in order that they could undergo extensive field tests.

The Quad and the Pigmy were both similar in many ways to the Blitz Buggy, but when it came to choosing which manufacturer they should go with, the army were unconvinced that Bantam would be able to supply their car in quantity in time. It was Willys who were awarded the final order for 16,000 vehicles, at about $740 each. Willys built their Quad incorporating design features from both the Blitz Buggy and the Pigmy, ultimately producing more than 335,000 of the new vehicles during the course of the war. The army used Ford's industrial might to build another 280,000. The Jeep was used in every theatre of war and proved itself to be a rugged and reliable vehicle.

And the famous name? Some say that Jeep came from the initials GP, for the designation 'General Purpose' vehicle, while others say that GP was actually a Ford production code – G for 'Government' and P identifying the type of chassis. There's another theory that soldiers named the vehicle 'Jeep' after Popeye's alien cartoon character friend Eugene the Jeep. Eugene had magic abilities and could get Popeye out of tricky situations.

However it acquired its name, Jeep is now a car brand that is part of the Chrysler corporation – and the name used by some of us for practically any car that can charge across a muddy field without you having to get out and push.

THE COMING
OF THE JET AGE

(THE JET ENGINE)

Most people believe that the invention of the jet engine came about around the time of the Second World War and, if we ignore the curious experiments with steam nozzles conducted by the ancient Greeks 2,000 years ago, most people are right. A practical jet engine intended for use in an aircraft was developed during the Second World War by the Germans and the British (obviously independently, not working together) for use in fighter aircraft, although the concept of the jet engine had been around for far longer.

A jet engine basically sucks air in through the front and then pushes it out the back, propelling the engine forwards, taking whatever it is attached to along for the ride. A spinning propeller does much the same thing, in water or in air, the shape of its blades carving through the air or water to push it out behind and provide forward motion. With a jet engine, however, the flow of air is controlled, it is compressed or squeezed under immense pressure, then mixed with fuel and ignited. The air in front of a jet engine is still and cool, but by the time it comes out the nozzle at the rear of the engine it is hot and travelling very fast, having been involved in a kind of explosion inside the engine, the power from which is all directed backwards to push the engine forwards.

The theory of how a jet engine should work had been understood for quite some time before anyone actually built one that worked. A working engine had to wait until the right kinds of metal had been devised and the engineers could produce components that would perform to the standards required. In 1910 a Romanian engineer named Henri Coandă designed an aircraft powered by a jet that used a conventional piston engine to suck in the air, compress it and expel it, although there is no real evidence that his aircraft ever actually flew. As there did not appear to be any facility for injecting fuel into the air to provide the explosive power of what we know as a jet engine, it is unlikely to have had enough thrust to get off the ground and was probably more like a giant hair dryer.

Twenty years later, a Royal Air Force engineering officer, Frank Whittle, applied for a patent on a jet engine he had designed. His patent was eventually granted but he had no real official backing for his project, developing it with the help of a private engineering company until he finally had a working prototype in 1937. At this point, the British government did start to take notice, especially since there were rumours coming out of Germany about a new aero engine being developed by an engineer named Hans von Ohain, with the backing of the Heinkel corporation.

Von Ohain's engine went through several development stages, at one time shooting massive flames out the rear end, much to the alarm of the technicians, but was working well enough by 27 August 1939 to power the experimental Heinkel He 178 into the air – the first flight of a jet aircraft. Following Ohain's limited success, a team was established at the Junkers aircraft company to develop their own jet engine, which they did with remarkable speed. It was this engine that powered the world's first jet fighter, the Me 262 A 'Swallow', when it took to the skies in July 1942. The aircraft entered service with the Luftwaffe in April 1944.

Just three months later, the Allies' first jet, the Gloster Meteor, entered service with the RAF, Frank Whittle's engine finally having been developed into a reliable powerplant. The Swallow and the Meteor never actually met in combat. The jet age, however, had arrived and the jet engine would be subject to continuous development. Within eight years commercial jet airliners were introduced and the world suddenly became a far smaller place.

CELEBRATING WITH A TWO-FINGERED SALUTE

(THE LONGBOW)

Bows and arrows have been used for hunting and in battle for thousands of years, the oldest archaeological discoveries that can be positively identified as bows dating back almost 12,000 years. The most significant type of bow developed for military use is a little more modern, the earliest English longbow having been discovered in Somerset and estimated to be almost 5,000 years old. It was much later, however, that the English longbow dominated the battlefields of Europe.

The longbow, made from a stave of yew around 6 feet (1.83 metres) in length was strung with a cord of hemp, flax or even silk. The string was attached to the bow using 'nocks', which were end caps for the bow tips notched with a slot to take a loop at the end of the string. This allowed for the bow to be easily unstrung when not in use, releasing the tension in order that it would retain its springiness. A typical longbow would fire an arrow about 30 inches (75 cm) long, with a metal arrowhead.

By the middle ages, the longbow archers of the English armies, many of whom were actually Welsh, were greatly feared. Their expertise was a decisive factor in battles such as Crécy in 1346 and Agincourt in 1415. Legend has it that prior to the Battle of

Agincourt, the French had threatened that they would cut off the index and middle fingers of any English archer they captured. When the massed longbows of the English drove the French back, the archers are said to have waved two fingers at the enemy, taunting them and establishing the 'two-fingered salute' that is still seen as an insult in many countries.

Sadly, although it's a great story, the 'two finger' legend is highly unlikely to be true. Any common archer captured would simply have been executed. Only noblemen would be held as prisoners, their families having enough wealth to be able to pay a ransom. Any other prisoners would simply be extra mouths to feed and that burden was best avoided by putting them to death, not cutting off their fingers.

The reason that the English archers were so feared was because of their discipline and training. Most had been required by their lords and masters to become expert archers, training for years as fathers and sons to build up the muscles and the stamina required repeatedly to draw and fire the longbow. The skeletons of archers from the period show that they had highly developed left arms and extra bone growth on their left wrists and shoulders as well as on the fingers of their right hands.

An archer could loose six arrows a minute, but would quickly tire if he tried to sustain that rate of fire. In battle conditions, he would be supplied with approximately 70 arrows and the massed ranks of archers – at Agincourt there were 7,000 longbowmen – fired on command. Firing in volleys, they could send a hail of thousands of arrows raining down on an advancing army at a range of up to 400 yards (365 metres). At 180 yards (165 m) a longbowman could hit an individual soldier. The closer the range, the greater were the chances that the arrow would penetrate a knight's armour, but even at the limit of the longbow's accurate range, all but the most robust, most expensive armour was vulnerable and the habit that the archers had of sticking their arrows into the ground at their feet, where they could get at them more quickly than they could arrows in a quiver, meant that the dirty tips caused nasty wound infections.

As firearms became more reliable and more widely used in the 16th and 17th centuries, the longbow became obsolete, although its silent action and accuracy meant that it still had its uses. During the Second World War, Lieutenant Colonel 'Mad Jack' Churchill led his men into battle armed with a Scottish claymore broadsword in his hand and a longbow over his shoulder. He is credited with at least one longbow kill, having used the weapon when he and his men ambushed a German patrol in 1940.

THE SHOCK OF
THE CENTURY

(SATELLITES)

For many, the greatest shock of the 20th century came on 4 October 1957 when the Soviet Union launched *Sputnik 1*, the first man-made satellite to go into orbit. In America, people were genuinely frightened by *Sputnik*, believing that the pesky Russkis now had the ability to look down from outer space, see everything every American was doing and drop bombs on the United States at will.

Sputnik, of course, could do no such thing. The satellite was a polished sphere of aluminium alloy, about the size of a microwave oven, with four antennae sprouting from its rear. It carried some basic scientific instruments, but the insides of the metal ball were mainly taken up with two radio transmitters and the batteries to power them. The bleeping signals from the radios provided valuable information to scientists about radio transmission through the atmosphere but they were also open signals, deliberately made powerful enough to be picked up by any radio operator. The Soviets wanted everyone in the United States to know that they had the technology to put a satellite into orbit. The polished surface of *Sputnik* was also purposely designed to reflect as much light as possible, making it highly visible. Away from the interference of street lights, Americans could look up into the sky and see the Soviet satellite passing over them. At least, they thought they could see it – what many were actually looking at was the far more visible final booster stage of the rocket that sent *Sputnik* into space.

The first satellite was a massive propaganda coup for the Soviet Union, a country that most Americans believed to be many years behind them in terms of industry and technology. At a time when, despite a temporary economic recession, Americans were cruising around in gargantuan, gadget-laden, gas-guzzling cars that struggled to achieve 15 miles to the gallon, were watching huge colour television sets years, if not decades, before most other countries had colour TV, and took refrigerators, air conditioning and the most modern domestic appliances very much for granted, it was deeply distressing for them to be made aware that they were not the most technologically advanced nation on the planet.

In fact, although the Americans had no way of knowing it, that probably wasn't entirely true. *Sputnik* was actually a far more basic satellite than the Soviets had intended to launch. Convinced that the Americans were going to beat them into space, they scaled down their ambitions and went for a design that would work, but that was less sophisticated than their original plans. They wanted to be first, and they wanted the Americans to know it. The frightening thing for military commanders in the West was that the Soviets sent *Sputnik* into orbit using the R-7 Semyorka rocket, proving that they had a missile that was capable of being launched in the Soviet Union to deliver a nuclear warhead anywhere in the world, including the United States.

The Americans, on the other hand, had a rocket capable of targeting the Soviet Union and capable of placing a satellite in orbit. They had been ready to do just that a year before *Sputnik* was launched, but had held back, unwilling to use a military missile system to put a satellite in space for fear of being accused of warmongering. Quite who amongst the general public would appreciate that the tried-and-tested Jupiter rocket created by Wernher von Braun's team at the US Army Ballistic Missile Agency was more military than the 'civilian' Vanguard rocket built by the US Navy is a mystery, but two months after *Sputnik* began orbiting the Earth, the less developed Vanguard system was chosen to launch America's first satellite. It rose to a height of four feet (1.2 metres) and exploded live on TV. The American press called it 'Kaputnik' and in the United Nations the sneering Soviets offered America help under a programme they ran to provide technical assistance to less developed nations.

That really fired the starting pistol for the race into space and the American government began pouring money into scientific research and the general education system, as well as giving the go-ahead for von Braun's team to launch a satellite using their Jupiter rocket, which they did in January 1958 with Jupiter masquerading as a civilian called Juno.

The American effort to overtake the Soviets resulted in the establishment of NASA (National Aeronautics and Space Administration) and the development of a mind-boggling array of technology from microelectronics and medical equipment to aeronautic advances and the internet. Since the 1950s thousands of satellites have been launched into space, with more than 50 countries sending hardware into orbit. Satellites are used for communications, navigation, scientific research and, of course, military skulduggery – spying on and taking photographs of other nations. *Sputnik* was not capable of doing what the American public feared it could do, but satellites exist today that most certainly can.

WE ALL LIVE
IN A ...

(THE SUBMARINE)

It may seem hard to believe, given that we tend to think of the submarine as a thoroughly modern invention, but the very first one is believed to have been tested almost 400 years ago.

A Dutch scientist named Cornelius Drebbel, working for King James I of England, constructed an underwater boat in 1620 that had been designed by an English mathematician some years before. It was a wooden structure covered in waterproofed leather that was rowed along underwater using oars. The first submarine was of more value as a curiosity than it was as a military vessel, but observers could not fail to see its potential for attacking enemy shipping.

By 1775 that potential was being put to the test when an American inventor named David Bushnell designed a one-man wooden submarine, a little like a giant wooden barrel, that was used in an attempt to drill holes in the hull of HMS *Eagle*, part of the British fleet blockading New York Harbor, and to place an explosive charge on the ship's hull. The attempt ultimately failed, but this was the first time that a submarine was used to attack a warship. Success in such an endeavour came in 1863 when another American sub, the 40-foot (12 metre), iron-hulled *H.L.*

Hunley used an explosive charge to sink the USS *Housatonic* during the American Civil War. The sub itself sank after completing its mission, killing the eight-man crew.

The *H.L. Hunley*'s success demonstrated that the submarine was a viable weapon of warfare and navies around the world began taking this new type of craft more seriously. While the *H.L. Hunley*'s propulsion had been via a propeller hand-cranked by the crew, new designs began to emerge that used steam power or electric motors and were armed with torpedoes, but by the beginning of the 20th century the familiar configuration of diesel engines for surface propulsion and electric motors for use when submerged became the norm. Submarines saw limited action in the Russo-Japanese War of 1904-05 but by the outbreak of the First World War in 1914, the sub was ready to show its true value as a strategic weapon.

Deploying their surface fleet was a real headache for the German Navy because the British Royal Navy was determined to keep the most powerful enemy ships bottled up in their home ports, denying them access to the Atlantic. The German U-boats (*Unterseeboot*, literally 'under sea boat') had less of a problem sneaking past the Allied patrols and could roam freely in the 'war zone' that had been declared around the British coast. The U-boat commanders developed tactics which included hunting in 'packs' to attack convoys and were responsible for the sinking of 5,000 Allied ships during the First World War, including the passenger liner *Lusitania*. The *Lusitania* was en route from America in May 1915, carrying a substantial cargo of munitions as well as almost 2,000 passengers and crew. More than half of them were killed when the ship went down, with 128 Americans among the fatalities, helping to turn the tide of public opinion in the United States in favour of America abandoning its neutrality to become involved in the conflict.

Between the First and Second World Wars, submarines were developed with greater range and more firepower but their role remained exactly the same and the country with the largest submarine fleet was Germany. The Kriegsmarine (as the German Navy was now known) U-boats were at their most effective in the Battle of the Atlantic, hunting convoys bringing vital supplies to Britain from the United States. They sank around 3,000 Allied ships, claiming the lives of up to 85,000 Allied seamen and sending millions of tons of food and supplies to the bottom of the sea.

Submarines became ever more sophisticated and grew ever larger during the course of the Second World War, acquiring radar for use on the surface, sonar for use underwater, 'breathing' tubes that would allow their diesel engines to be used while the boat was submerged and myriad other improvements. During the Cold War era, however, the submarine was to take on a new role. In 1953, the USS *Tunny* launched a missile that was capable of carrying a nuclear warhead and the submarine became a major player in the nuclear arms race. By the 1980s, this had led to the building of the world's biggest submarine, the Soviet *Typhoon* class. Six of these monsters, 574 feet long (175 metres) and as tall as a nine-storey building, were built. They were powered by nuclear reactors and could remain submerged for four months. Capable of passing undetected beneath the polar ice cap, they were designed to position themselves off the coast of America to launch their missiles. They were armed with 20 ballistic missiles, each of which had ten warheads that could be individually targeted and each of which had more than ten times the destructive power of the atomic bomb that destroyed Hiroshima during the Second World War. While no other country had a sub quite as big as the *Typhoon*, the United States wasn't too far behind with their *Ohio* class subs.

The underwater row-boat novelty built for King James had grown through four centuries to become a monster – fleets of monsters – prowling the depths of the ocean with enough firepower to destroy the world.

BATTLESHIPS FOR THE BATTLEFIELD

(THE TANK)

The Land Ironclads is what H.G. Wells called them in his eponymous story in 1903 when he wrote about an army of city dwellers facing an army from the countryside, each of them defending well-excavated trenches on a shell-blasted battlefield. The country soldiers, with their superior knowledge of fieldcraft, superior fitness and superior fighting skills, are convinced that they will win, but then the army from the city uses its superior technology by deploying the Land Ironclads – tanks. The country soldiers have no weapon that can stop the tanks, and the battle is lost.

One of the things that made H.G. Wells' fiction so popular – he was known, along with Jules Verne and Hugo Gernsback, as one of the 'Fathers of Science Fiction' – was that he gave what seemed like pure fantasy a firm grounding in contemporary science and engineering. An 'ironclad' was a steam-powered battleship protected by iron or steel armour plating. It didn't tax Wells' highly fertile imagination too much to transfer the concept of a heavily armoured fighting vessel from the ocean onto the battlefield, and even as he wrote the story, military men were laying plans to do exactly that.

The Land Ironclads moved, as Wells described them, 'on eight pairs of big pedrail wheels, each about ten feet in diameter'. The pedrail was a wheel equipped with numerous spring-loaded feet around its rim which allowed it to grip a soft, muddy surface and, when used in the way that Wells described, the pedrail could move a vehicle across uneven ground without it bucking, rolling or shaking itself to pieces. Invented by London engineer Bramah Joseph Diplock, the pedrail was considered for use in the first tanks but was rejected in favour of the simpler, more robust and less expensive caterpillar tracks. Military minds were working way ahead of the science fiction writer's imagination in finding a practical way to bring a battleship onto the battlefield.

Although the concept of the tank seems like an obvious idea to us in the 21st century, it wasn't such a simple thing to create at the beginning of the 20th century. The motor car was very much in its infancy, the internal combustion engine still to be refined to a stage where it could be considered sufficiently reliable to power a battlefield mobile offensive weapons system. Progress on the development of such a system was, therefore, as painfully slow and prone to unexpected breakdowns as the motor vehicles of the time.

In 1903 a French artillery officer, Captain Levavasseur, proposed mounting a heavy field gun on an armoured chassis that would be propelled by its own engine, its wheels running on continuous crawler tracks to spread the weight of the machine and stop it sinking into soft ground. His idea was eventually rejected by his superiors who thought that horses could haul artillery pieces just as well. Levavasseur had actually invented the tank fully 13 years before one ever went into battle. One of the problems was finding a crawler track system that would work reliably. American inventor Benjamin Holt was one of those working on the idea. In 1903, after visiting Britain, which had become the centre for such track research, he started working on his own system, buying and developing patents from American engineer Alvin Lombard and British company Hornsby to create a working tracked vehicle, intended primarily for agricultural use. The Holt tractor was tested by the military for hauling artillery, at which point a British soldier apparently joked that the machine moved like a caterpillar. Holt adopted the term and his company ultimately became Caterpillar Inc., now an instantly recognizable brand and a household name worldwide.

By the time the First World War had settled into the misery and stalemate of trench warfare, it was clear that an armoured vehicle capable of crossing rough ground and trenches was required to break the deadlock. The French experimented with a wheeled steel monster based on an industrial machine originally used in canal building, the Frot-Laffly Landship, while the British efforts became more focused in 1915 when Winston Churchill, then First Lord of the Admiralty, formed the Landships Committee. At this point it looked like the Royal Navy might take to the battlefields of Europe in land battleships (hence the interest from the Admiralty) and, although the army eventually gained control of the vehicles under development, naval terms such as hatch, bow and hull endured. The prototypes included pedrail systems, wheeled vehicles and tracked designs. All failed to perform adequately in trials but it was the tracked version that warranted further attention, with revised track designs and a rhomboid configuration beginning to show promise. The end result was called a 'tank' to hide its true identity. The documentation described the vehicles as water carriers and the engineers working on their construction referred to them as 'water tanks'. The word 'tank', sounding innocuous enough to conceal their true purpose, was soon adopted by everyone. Even Churchill's Landships Committee changed its name to the Tank Supply Committee.

The French had adopted the Holt tractor system and were developing their own tanks at the same time as the British, although it was the British Mark I tank that first saw action. It was not a complete success. Almost 50 tanks were to be deployed during the Battle of the Somme on 15 September 1916. A third of them suffered mechanical problems and took no part in the action. Many broke down or became bogged down on the battlefield but enough were able to make it across no man's land at their cruising speed of around 4 mph to make a significant contribution. They destroyed barbed wire, crossed trenches and even bulldozed houses where machine-gun nests were hidden.

The German troops who saw the British Mark Is coming towards them must have been terrified. These rhomboidal machines, a kind of squashed diamond shape with crawler tracks running all the way round the edge of the hull on each side and the crew compartment huddled between the crawlers, were armed with either machine guns or cannon. These were mounted in turrets sticking out from each side of the tank. The steel monsters seemed impervious to rifle or machine-gun fire. A direct hit from a mortar or an artillery shell could ignite the fuel tank and destroy it, or break a track and disable it, but when they first appeared, the tanks must have looked invincible to the Germans. On the inside, however, it was a different story.

The early tank crews shared their working space with the engine, deafened by the noise and choking on fumes. Cordite from the machine guns and cannon added to the foul air as did the stench of the eight-man crew, working in darkness and temperatures of up to 50° Celsius. They were also obliged to wear helmets and padded leather face masks armoured with steel 'chainmail' to protect them from flying shrapnel when bullets hitting the outside of the tank caused rivets or slivers of metal to ping off the inside. Tank crews were known to pass out for lack of air or, when they stepped out of their vehicle, to faint as soon as they hit the fresh air.

Despite their shortcomings, tanks were soon in regular use on the battlefield, with the Germans using captured British tanks and the innovative French developing a range of tanks that included the Renault FT light tank. The little Renault had a crew of just two but established the layout that all tanks were to follow. Its driver sat in front, the engine was in the rear and the commander was ensconced in a revolving gun turret on top. It was small by the standards of other tanks of the time (the German A7V was over 24 feet long and had a crew of 18) but the Renault remained in service with more than two dozen different countries well into the Second World War.

Between the wars, tank development slowed as governments slashed their military budgets but tanks, just like motor cars, became more reliable and different military strategies evolved using tanks to promote mobile warfare and avoid the stalemate of the trenches. The Germans showed their mastery of mobile strategy with their Blitzkrieg tactics at the outset of the Second World War, although their early Panzer tanks were outperformed by others such as the Soviet T-34. Tanks were used in every theatre from the North African desert to the tropical forests of the Far East but it was in eastern Europe that the biggest tank battle took place when 3,000 German tanks faced 8,000 Soviet machines in a series of clashes between July and August 1943 that came to be known as the Battle of Kursk. Thousands of artillery pieces and aircraft as well as millions of men were involved, with the Soviets ultimately driving their enemy back towards Germany.

At the end of the Second World War, with Europe divided and the West facing a massive arsenal of Soviet firepower, it was envisaged that when the 'Cold War' turned hot, tanks of the

Warsaw Pact nations would sweep westwards. By the beginning of the 1980s, it was estimated that NATO countries could field around 30,000 tanks, while the Warsaw Pact could call on more than twice that number. These modern tanks were a far cry from the juddering monsters of 1916. They could be sealed to protect the crew from nuclear fallout or biological weapons; they had sophisticated radio and radar systems; their main guns had computer-controlled targeting systems; they had infra-red and night vision systems to spot their enemy in the dark; they had far more effective armour; and some were capable of achieving 40 mph rather than the First World War tanks' 4 mph.

Thankfully, there have been no European tank battles since the end of the Second World War but the tank has seen service in every major conflict since, including the wars in Iraq and Afghanistan. Modern tanks have armour that First World War tank crews could never have imagined. Special ceramics combined with metal compounds provide high-tech protection along with 'reactive armour' which can explode to deflect the effects of armour-piercing shells. The British Army's Challenger 2 tank proved the value of its armour in Iraq in 2003 when one Challenger was attacked in the city of Basra. The tank was hit by heavy machine-gun fire, an anti-tank missile and 14 rocket-propelled grenades. The crew were able to return fire until their sighting systems were damaged and they reversed into a ditch, losing a track. Then they just sat tight, safe inside their tank, the armoured hull taking the punishment until a battlefield recovery crew could reach them. The Challenger was operational again within six hours.

H.G. Wells predicted that Land Ironclads would change the tactics of the battlefield and tanks certainly have altered the nature of warfare, changing the world in the process. There is little doubt that they will continue so to do. Tanks that are currently on the drawing boards include unmanned machines that are not operated by remote control but that can do battle independently, thinking for themselves, identifying their enemies and engaging them as they see fit. In sci-fi terms, concepts like that take a giant leap forward from H.G. Wells towards the *Terminator* movies and the rise of the machines!

TALKING ON THE GO

(THE WALKIE TALKIE)

We are all used to seeing 'walkie-talkies' being used today. They have become a part of everyday life. The emergency services use portable personal radios; they are used on construction sites; sports coaches use them pitchside to talk to managers or trainers; and a modern walkie-talkie is small enough to slip into your pocket to use on a day out when parents are playing with, or trying to keep tabs on, their kids.

Using a radio to talk to someone – even someone nearby – hasn't always been that easy. Portable two-way radios with voice communication are taken very much for granted in the 21st century, but it wasn't until the Second World War that a practical system was developed.

In the late 1930s a British-born inventor, Donald Hings, who had lived in Canada since he was a child, devised a portable radio for use by bush pilots operating in remote areas of the Canadian wilderness. Hings' radio was simply known as a 'pack set' and no one really took too much notice of it. Then, along came the Second World War and military men were suddenly very interested in Hings' device. The radio was adapted and refined for military use, with thousands being produced to be mounted in vehicles and tanks as well as carried, papoose-like, strapped to the chest of an infantry signaller.

It's not entirely clear whether the phrase 'walkie-talkie' was first applied to Hings' radio or to the SCR-300, produced by Motorola for the US Army Signal Corps. With none of the miniaturised electronic components that are packed into modern gadgets having been invented, the Motorola engineers had to incorporate 18 glass vacuum tube triodes (transistors would make these obsolete within the next ten years) into a radio pack that was robust enough to survive on the battlefield. Their SCR-300 was carried in a backpack and weighed around 35 pounds (16 kg) but was a rugged, waterproof radio that the army found hugely impressive.

From the time that Motorola were briefed by the army on their requirements in late 1940, it took only six months to develop a working prototype. The army urged Motorola to forge ahead in refining the SCR-300 for manufacture. A total of 50,000 units were eventually produced, their first major battlefield test coming during the Allied invasion of Italy in 1943.

Hings' pack set and the Motorola back pack set were dubbed 'walkie-talkies' because, of course, you could walk while talking. Another breakthrough in this nascent mobile technology came with the 'Handie-Talkie'. Alongside the SCR-300, Motorola developed the SCR-536, known as the Handie-Talkie. This was a radio transmitter/receiver that weighed only 5 pounds (2.3 kg) and was incredibly small. If it was a box, it could have held a couple of house bricks. The hand-held unit had a more limited range than the backpack radio, but it was ideal for the battlefield, Motorola producing 130,000 of them during the course of the war.

Motorola actually trademarked the Handie-Talkie name but it never caught on in the same way that walkie-talkie, a name that was coined by journalists and soldiers, had done. The company continued at the forefront of mobile communications, with the head of Motorola's systems division, Dr Martin Cooper, making the first call from a hand-held mobile telephone in 1973. Cooper became known as 'the father of the cell phone', although the device he used in 1973 resembled the Handie-Talkie, being about the size of a house brick, much more than it did a modern mobile phone.

PUTTING THE FRIGHTENERS ON GOLIATH

(THE TREBUCHET)

There was a time when, if you spotted an enemy army approaching, you could call all the local villagers inside the walls of your castle and sit tight, having prepared for a long siege by storing enough food to keep you going for months, keeping enough livestock within the castle walls to top up your supplies and raising water from a well deep below ground that was impossible for the fiends outside to poison. If they came near your castle walls or tried to batter down the gate, your archers could pick them off from the ramparts or you could pour something unpleasant like boiling oil down onto them.

The likelihood was that the attacking army would run out of food or patience before you did and trundle off to bother someone else, somewhere far enough away for you not to care too much about it. For the attackers, breaking into a castle was a troublesome business ... until the trebuchet came along.

The trebuchet is one of a number of devices that are described as 'siege engines' and, when compared to the protective sheds on wheels with a battering ram underneath to attack a gate or giant wheeled-tower staircases that could be rolled up to the castle wall, the trebuchet was the most dynamic. It was a true engine in the sense that it converted power into movement but in skilled hands it was also a devastating artillery piece.

Essentially the trebuchet was a huge slingshot, working on the same principle as the one that David used to bring down Goliath. To use his slingshot, a patch of leather that had a thong or string attached to either side, David gripped the ends of the string together in one hand with a stone nestling in the now folded leather patch. He whirled it around at arm's length and then flung it towards Goliath, letting go of one of the strings. The stone shot forward and smacked Goliath right between the eyes, felling the giant warrior on the spot.

David's basic slingshot became slightly more sophisticated over the years with the addition of a stick that extended the user's arm length, providing a greater lever for the final throw to impart greater velocity and better range to the missile. By the 5th century BC, the slingshot or catapult concept had been extended even further in the shape of a machine that worked a bit like a see-saw. A heavy boulder was loaded on the end on the ground, then a squad of men hauled down on ropes attached to the other end, causing the boulder to fly through the air. A boulder, or a barrage of such boulders from a formation of machines, could decimate an approaching cavalry formation or reduce the walls of a town to rubble. This was the earliest form of trebuchet.

Other kinds of long-range weapons were developed. The Romans used the onager, a kind of catapult that relied on twisting rope to provide spring-like tension for propulsion. They also had the ballista, a kind of giant crossbow that launched huge arrow missiles. Like the trebuchet, these were developments of earlier technologies, but the later type of trebuchet was capable of delivering far heavier projectiles with great accuracy.

The counterweight trebuchet didn't rely on the strength and co-ordination of a squad of men. Its see-saw power came from huge weights attached to increasingly sophisticated hinged pendulum arrangements which used gravity to pull the throwing arm into the air. A slingshot, laid out along the ground in readiness, then whipped forwards to send the missile thundering towards the target. By the 15th century, rocks weighing hundreds of kilograms could be hurled up to 300 metres and there are reports of some trebuchets in the Far East slinging rocks weighing more than a ton.

In 1304, Stirling Castle in Scotland was besieged by the English King Edward I, who brought with him a number of siege engines, including a trebuchet called *Warwolf*. Today, *Warwolf* would come with an advisory note warning that 'Some self-assembly is required.' Its component parts were transported in 30 wagons and it took a team of carpenters three months to build. In the meantime, the Scots didn't much like what was going on outside and, to Edward's great disappointment, surrendered. Having gone to all the trouble of building *Warwolf*, he wasn't about to pack it all away again without slinging a few stones, so he sent the Scots back inside and opened fire, destroying part of the castle wall and gatehouse.

The trebuchet was quite versatile when it came to choosing your ammunition. You could hurl an assortment of rocks, clay balls that would burst open and spill out smouldering debris that set fire to wooden buildings within the castle walls, and in 1422 during the siege of Karlstejn Castle in what is now the Czech Republic, corpses and dung were flung. This caused disease to spread within the castle – an early form of biological warfare.

Trebuchets continued to be used long after gunpowder and cannon made their appearance on the battlefield. Their range and accuracy made them as effective as cannon against castle walls and they could be used repeatedly without overheating the way that cannon tended to do. They made their presence felt in conflicts as far afield as China, the Middle East, Scandinavia and Europe, often turning the tide of battle in much the same way that David had with his slingshot.

THE GUN
THAT WON
THE WEST

(THE WINCHESTER '73)

Hollywood stars James Stewart and Shelley Winters fought the bad guys, bad luck and bad Indians (including Rock Hudson as 'Young Blood') with the help of a trusty Winchester in Universal Pictures' 1950 movie *Winchester '73*. It opened with a caption on screen that read: 'This is the story of the Winchester Rifle Model 1873, "The Gun That Won The West". To cowman, outlaw, peace officer or soldier, the Winchester '73 was a treasured possession. An Indian would sell his soul to own one.'

That statement, perhaps, better reflects the attitudes of the time towards people that we now more respectfully refer to as Native Americans, who undoubtedly valued their souls rather more highly than even a Winchester which, around the time that the '73 was introduced, cost about $20 – a month's pay for most ranch hands. It was an expensive firearm, one that cowboys and settlers in the old west really did treasure, and that Native Americans certainly coveted. The Winchester had the stopping power to take down a bear, a mountain lion – or, unfortunately, an 'Indian' – at a range that allowed you to take several pot shots at your attacker before it – or he – was upon you. And the famous 'lever action' meant that you could reload and fire very quickly, with a tubular magazine beneath the barrel supplying up to 15 rounds for you to nail the pesky varmint.

The 1873 model was a development of the original Winchester Rifle, the 1866, but its story goes back even further than that. Oliver Fisher Winchester was a successful clothing manufacturer in New York City and New Haven, Connecticut. In 1857 Winchester took over the Volcanic Repeating Arms Company, formerly run by Horace Smith and Daniel Wesson who left to form the Smith & Wesson Revolver Company. Volcanic's main product, the Volcanic Rifle, had not been a huge success, but Winchester saw great potential in its innovative lever-action design. He renamed the company the New Haven Arms Company and had its leading gunsmith, Benjamin Henry, redesign the rifle and the ammunition it used to make it more efficient and more reliable. In 1860, New Haven launched the Henry Rifle, which was used by some Union Army soldiers during the 1861-65 American Civil War. Most troops during that war were still using single-shot, muzzle-loading rifles or muskets that were little more than hand-portable cannons with a rate of fire of only about three shots per minute. The Henry Rifle could deliver 28 rounds per minute. Although it wasn't widely issued to Union soldiers, those who could afford the $40 price tag of a Henry Rifle bought one because they knew it might just save their lives. On the other hand, the Confederate troops who came under fire from the Henry described it as 'that damned Yankee rifle that they load on Sunday and shoot all week!'

When the New Haven Arms Company became the Winchester Repeating Arms Company in 1866, that lever action (where pushing down on the handle that was integral to the trigger guard and then pulling it back up into place drew a round from the spring-loaded magazine, locked it into the chamber and cocked the weapon ready for firing) became synonymous with the name Winchester, despite other arms manufacturers using similar lever systems. The

Winchester '86 was put to good use by Turkish troops against Russian forces at the Siege of Plevna with vastly outnumbered Turks holding the city through three battles before being overrun.

It was the Winchester 1873, however, that guaranteed the rifle a place in history with more than 720,000 of the model being produced. Yet 'The Gun That Won The West' was not widely used by the US Army. The Canadian North-West Mounted Police used the '76 model, as did the Texas Rangers, and other derivatives were sold to armies around the world, but the US Army decreed that the Springfield Rifle, a single-shot weapon, should be issued to its troops. They believed that the Springfield was more reliable, more accurate and more powerful. At the Battle of the Little Bighorn, General Custer's 7th Cavalry troopers were using Springfields while many of the Cheyenne, Lakota and Arapaho warriors had Winchesters. Custer's men were massacred. Ironically, in 1955 Colombia Pictures stole Universal's 'Gun That Won The West' catchphrase for their eponymous film about the Springfield.

The '73 was followed by the '76, the '86, the '92 and other new, improved designs, all of which, to the untrained eye, looked very similar and all of which were commonly known as Winchesters. They could be ordered with smaller magazines, special finishes on the wooden buttstock and forearm or, as in the case of President Theodore Roosevelt's hunting rifle, elaborate engraving on parts of the metalwork. John Wayne used Model 1892 Winchesters in many movies, regardless of what year the action was taking place, and owned several himself, with Winchester announcing a special Model 1892 John Wayne 100th Anniversary Rifle in 2006 – in time for the centenary of John Wayne's 1907 birth.

The Winchester 1894 became the best-selling hunting rifle of all time with seven million produced up until 2006 when the New Haven plant was closed down. The Winchester Repeating Arms

Company is now part of a much larger, modern arms manufacturer but The Gun That Won The West continues to be made in small numbers under licence by companies in Brazil and Italy, while Winchester occasionally produces special editions, such as the John Wayne model.

GOODBYE
MR GOODBAR

(THE MICROWAVE OVEN)

It's the sort of thing that could happen to any small child. They put a bar of chocolate in a trouser pocket, looking forward to having it later, and before they know it, after a few minutes of rushing around, it's melted all over the place. It happened to Percy Spencer in 1945 when a 'Mr Goodbar' Hershey bar made a fine old mess of his trousers, but Percy hadn't been rushing around playing with his friends. He was a 51-year-old engineer in charge of 5,000 people and wasn't getting overheated. The chocolate shouldn't have melted.

Percy Spencer was in charge of a workforce producing the magnetron. As a leading expert on the design of radar sets, Spencer had revolutionised production of the magnetron, a British invention that produced high-power pulses of microwave energy. The magnetron allowed radar sets able to detect far smaller objects and for the sets themselves to be made small enough for use on aircraft. Spencer worked for the Raytheon company in the United States, who had taken on and, under Spencer, vastly improved the manufacturing process of the magnetron, producing them more quickly. During the Second World War, this had been a major achievement with radar systems having been given priority in terms of research and development that was second only to the Manhattan Project nuclear weapons programme.

Spencer was working on a magnetron one day, standing by an active radar set, when he had the unfortunate accident with his chocolate. The microwave emissions from the radar set had melted the chocolate in his pocket. While others had noticed this strange effect before, no one had paid it much attention. Spencer's highly active mind, on the other hand, needed to know more about why this was happening and how the effect could be harnessed.

He worked out how to construct a closed metal box into which he could direct magnetron microwaves, the metal box giving them no escape route. Placing food in the box, it was found that the energy from the microwaves penetrated the food, agitating the food molecules in such a way that it caused the food to heat up. The first thing Spencer and his team tried it with was popcorn. The second was an egg, which exploded.

Raytheon developed the idea of a microwave oven, testing it in a commercial restaurant kitchen before offering the 'Radarange' oven for sale to the general public in 1947 – or at least offering it to the general public who could afford the $5,000 dollar price tag – probably around $50,000 at today's prices. Anyone who wanted one would also have to have a reasonably big kitchen as the Radarange stood almost 6 feet (nearly 1.83 metres) high. Part of the reason for its size was that it had to incorporate a water cooling system to prevent it overheating.

Like most other advanced devices, technological improvements over the years led to the microwave oven becoming smaller and less expensive until, by the mid-1970s, more than a million microwave ovens were being sold each year in America. Only 20 years later, 90 per cent of American homes had a microwave and that general pattern was repeated around the world – all because poor Spencer melted Mr Goodbar.

HARRY POTTER SYNDROME

(SCRATCH-RESISTANT LENSES)

If you wear glasses you will know the frustration of putting them down somewhere and not being able to find them. You search high and low, struggling to see them because, of course, you're not wearing them. Then, when you do find them, you reach out to pick them up and, because you haven't got your glasses on, you misjudge it, knock them onto the floor, step forward to retrieve them and ... crunch.

It seemed to happen to Harry Potter constantly. Those glasses of his were always getting cracked or smashed, usually just when he needed them most. It's the Harry Potter Syndrome. Obviously if Harry had had plastic lenses in his glasses he wouldn't have had such a problem. Using glass in spectacles is very old-fashioned nowadays because it's simply not safe. Glass will crack and splinter or shatter in a way that plastic will not. Plastic is also far lighter, meaning that spectacles (you can't really call them glasses if they're plastic) are more comfortable to wear. Manufacturing lenses out of plastic is also far cheaper, making your spectacles less expensive, and plastic is even better at absorbing ultraviolet radiation, giving better protection for your eyes.

So, all Harry had to do was to hop on his broomstick and shoot down to his nearest optician to pick up some spectacles with plastic lenses. There is, however, a problem with using plastic. It can easily become scuffed or scratched when used in the lenses of spectacles, making it difficult for you to see through them – not an ideal situation for Harry if he is zooming around playing Quidditch or he has a Death Eater or a Dementor sneaking up on him.

The solution to the scuffing and scratching is to have plastic lenses with a scratch-resistant coating, giving all of the advantages of plastic with the smooth, hard surface of glass. That is exactly what NASA was looking for to use on, amongst other things, the visors of astronauts' helmets. There can be a lot of dust and grit floating around in space, especially if you are bouncing around on the surface of the moon, and at NASA's Ames Research Center in California they came up with a solution while working on a coating for a membrane in a water purifier. The system used to apply the ultra-thin layer was later developed to create a thin, tough coating for helmet visors.

At NASA's Lewis Research Center in Cleveland, they devised a way to give plastic a coating that was as tough and scratch resistant as the hardest material in the world – diamond. The Diamond-Like Carbon (DLC) coating is now used on plastic lenses for high-quality spectacles and sunglasses but it also has applications when used on steel and other materials, not only in the aerospace business but to make a host of things from multi-blade razors to parts for artificial human heart pumps.

The next time you see a clip from a Harry Potter movie when his glasses get trashed, you'll know that what he needs is a bit of DLC – or maybe he could just use magic?

A GOOD NIGHT'S SLEEP

(MEMORY FOAM)

You wouldn't think that having a nice, restful snooze had much to do with falling backwards from outer space towards a crash landing at 25,000 mph, but scientists and engineers working on the crash-landing problem ultimately contributed towards the good night's sleep enjoyed by millions.

The Apollo Command Module, having carried its three astronauts to the moon and back, arrived in Earth's atmosphere travelling at roughly 25,000 mph. After the relatively stress-free journey through the vacuum of space, the Command Module – now serving as the Re-entry Vehicle – needed all the insulating properties of its heat shield to protect the occupants from the enormous temperatures generated by friction as the craft came into contact with the molecules present in our atmosphere. The friction also slowed the Re-entry Vehicle and, once it had fallen to a height of around 11,000 feet (3,333 metres), parachutes were deployed to slow it even further, although it still hit the ocean at about 20 mph. Various built-in impact zones absorbed the force of the capsule hitting the surface and some of the lessons learned about that impact absorption were later put to good use for other purposes.

Charles Yost was one of the aeronautical engineers who worked on the Apollo Command Module and he was later tasked with designing an aircraft seat that would protect the occupant in the event of a heavy landing. Yost created a plastic foam that was unusually elastic, allowing for high energy absorption while remaining soft enough to be comfortable. The open-cell, polyurethane-silicone foam was developed into a material that, even when compressed to 10 per cent of its normal thickness, would always bounce back into shape. NASA's Ames Research Center used the material in an aircraft seat design and it was also adopted for commercial aircraft seats.

The fact that, when you sat or lay on it, the foam would compress to fit the contours of your body, then gently return to its original shape when you got up, also made it an ideal choice for making mattresses. Bedridden patients in hospital saw enormous benefits from the 'memory foam' in a mattress distributing the weight of the body more evenly, helping to prevent bedsores.

The foam is also used in fitting amputees with artificial legs as it provides an ideal cushion between the existing leg and the prosthetic. It has been used in motorcycle saddles, racing-car seats and shoe insoles. Continued development and improvement of memory foam has even brought it back to NASA, who have used it as flooring for an obstacle course. Astronauts who have been floating around in the zero-gravity environment of space for long periods of time need to work hard to re-adapt to Earth's gravity. To test how well they are progressing in coping with the stresses of being a normal weight again, and in regaining their balance, they are sent over the obstacle course with a memory-foam cushioned floor.

For millions, however, what the invention of memory foam means is a good night's sleep.

THE STORY OF FLIGHT GOES VERTICAL

(THE HELICOPTER)

Helicopters are a familiar sight over most major cities. They are a convenient mode of transport for those wealthy enough not to want to waste their time in traffic jams or wait for trains along with the rest of us, but they are also vital to the emergency services. Their military applications, of course, have been exploited ever since the first helicopter took to the skies. Every military commander wants to have 'eyes in the sky' to be able to see what the enemy is up to and helicopters, like all other man-made flying machines, give them exactly that.

The British princes, William and Harry, both serve in the military as aircrew manning helicopters, although their jobs demonstrate the range of tasks that the helicopter can undertake. Prince William flies an RAF Sea King search-and-rescue helicopter, not only going to the aid of military personnel, but also providing an invaluable emergency service for anyone in difficulty at sea, lost or injured in the mountains or otherwise in trouble in a location not easily accessible to any other form of transport. Prince Harry, on the other hand, flies in an Army Air Corp Apache gunship armed with some of the most sophisticated weaponry and battlefield technology on the planet.

The helicopter, it would seem, is an enormously versatile aircraft, capable of doing all sorts of things. The world's largest helicopter, the Russian Mi-26 Halo, was designed to transport armoured vehicles and ballistic missiles, but in a civilian role was once used in Siberia to lift a block of mud and ice weighing around 23 tons. The frozen payload contained the remains of a woolly mammoth! Saving the mammoth for scientific study was perhaps the most bizarre search-and-rescue mission ever undertaken by any helicopter and only the most modern, powerful machine could have managed it – although even the mighty Halo struggled to lift such a massive load – but the principle of how a helicopter lifts anything, including itself, off the ground, is not a new idea.

The earliest mentions of a helicopter-type device come from China around 400 BC when children played with helicopter toys made from lightweight bamboo. The toy consisted of a vertical rod with two thin wooden blades fitted across the top. When the children spun the rod by rolling it between the flattened palms of their hands, the helicopter rose into the air. Fanciful Chinese stories told of flying cars taking passengers soaring into the heavens but, in fact, at that time there was no way of supplying the kind of power that would be required to turn blades capable of lifting anything.

The Chinese toy turned up in Europe in the middle ages, as did so many things from China, and Leonardo da Vinci sketched an idea for a helicopter-like flying machine with a revolving spiral blade, but the concept remained no more than an entertaining novelty, despite the toy being developed using springs and rubber bands that prolonged its flight. No one really even knew what to call these things until French inventor Gustave de Ponton d'Amécourt used the word 'helicopter' to describe the small, steam-powered model that he had built in

1861. The word came from the Greek *helix*, meaning 'twisted' and *pteron*, meaning 'wing'.

By that time, men had already ascended into the skies using kites (another Chinese invention) and hot air balloons (again Chinese). The first manned ascent in a free-floating balloon took place in France in 1783 but balloons were notoriously difficult to control, tending to go wherever the wind blew them, and a mechanical device that could soar straight up into the air was every aviation enthusiast's dream. By 1878, Italian engineer and aviation pioneer Enrico Forlanini used a steam engine to send an unmanned helicopter model up to a height of around 40 feet (12 metres), and by the turn of the century a number of would-be helicopter pilots were experimenting with designs using internal combustion engines. When the Wright brothers made their famous first powered flight in 1903, however, it was 'fixed-wing' aircraft that grabbed all the attention.

Why the term 'fixed wing'? Because an aircraft such as the Wright brothers' kite-like flying machine used the airflow over a wing, or wings, to provide the lift that made it fly. The aircraft has to be moving forwards at speed for the wing to achieve sufficient lift to get the whole thing airborne. A helicopter can be referred to as a 'rotary wing' aircraft. Its wings achieve the speed they need to generate lift by spinning through the air rather than running forwards. This does create a troublesome torque effect whereby, once a helicopter has left the ground, the spinning of the blades will start to turn the body of the aircraft in the opposite direction. Much the same effect is found in fixed-wing aircraft driven by a propeller in the nose. The spinning propeller will induce the whole aircraft to turn in a roll along its length once it is no longer braced on the ground. An aeroplane has to be 'trimmed' using its control surfaces to counteract this effect. A helicopter counteracts the effect of torque by using a tail rotor or having two rotors revolving in opposite directions.

Helicopter pioneers came to understand the problems and developed a number of ingenious solutions. In 1924 another Frenchman, Etienne Oehmichen, set the first helicopter flight record in an aircraft that used four rotors, managing to fly a distance of almost half a mile. The US Army Air Service also experimented with quadrotor aircraft around this time, although they ultimately abandoned their project, perhaps because Argentinian aero engineer Raúl Pateras Pescara de Castelluccio, broke Oehmichen's record, also in 1924, with his helicopter that used multiple rotors all spinning on the same axle above the aircraft. Pescara enjoyed funding from both the French and British governments and his machine could stay aloft, at about head height, for up to ten minutes.

Dutch, American, Hungarian, Russian and Italian enthusiasts all contributed their ideas over the next few years, slowly developing the technology that would make the helicopter a viable proposition, including autogyros that were part aeroplane, part helicopter. Then, in 1938, German pilot Hannah Reitsch, the only woman ever awarded the Iron Cross First Class, demonstrated the Focke-Wulf Fw61, the world's first practical helicopter. Reitsch's demonstration of Heinrich Focke's design took place indoors, at the Deutschlandhalle sports stadium in Berlin. The Fw61 looked like an aeroplane without wings, which is precisely what it was, having been built using the fuselage of a training plane with the engine in the front driving a propeller as well as two large, three-bladed rotors held on outriggers sticking out from the body roughly where the wings would have been. Proving that it wasn't just an indoor circus act, the Fw61 showed that it could really fly when put through its paces outdoors, demonstrating a range of 143 miles and the ability to reach an altitude of over 11,000 feet (3350 metres).

Nazi Germany was the first combatant to fly helicopters during the Second World War, using them for observation, casualty evacuation and transport, although the aircraft were small and only built in limited numbers with Allied bombing raids on their factories ensuring that they could not be mass produced. Because they could land and take off vertically, requiring only a small patch of ground as opposed to an aeroplane's runway, no one wanted the Nazi helicopters to become a tangible threat. In the meantime, Russian-born American aero engineer Igor Sikorsky was working on the XR-4 for the US Army Air Force. This was the first aircraft that we can really look at nowadays and immediately recognise as a helicopter. It had a single, three-blade rotor above the fuselage and a smaller vertical rotor in the tail to counter torque spin. The Sikorsky R-4 went to war in 1944, seeing service as a rescue helicopter and being used for casualty evacuation.

From the end of the Second World War, the helicopter's reliability and usefulness grew until it established itself as a vital piece of military hardware. They were widely used in the Korean War (1950-53) to carry loads and evacuate casualties and during the Suez Crisis in 1956 British Royal Marines became the first troops to be ferried into combat in an air assault.

Helicopters were certainly no longer novelty flying toys and as technology advanced they transcended the military, enjoying civilian roles as air ambulances, aerial TV and film camera platforms, forest fire fighters and executive transport, but without the military the helicopter would never have got off the ground.

A CLOUD
WITH A
SILVER LINING

(MUSTARD GAS)

Chemical warfare has been practised for centuries in one form or another. Poison-tipped arrows constitute a kind of chemical warfare and even the idea of poison gas is not unique to the modern world. Burning various substances in an attempt to incapacitate the opposition with noxious smoke has been tried many times throughout history but the indiscriminate nature of poison smoke or gas – the wind could easily blow it at your own troops – made it an unreliable weapon.

That was to change during the First World War when an effective delivery system, the artillery shell, could carry a payload of poison gas far from friendly troops. The wind could be judged with reasonable accuracy and, perhaps most important of all, the means to manufacture chemical agents on an industrial scale had been devised. The German Army first used mustard gas on British troops at Ypres in Belgium in 1917. It was known as mustard gas because of its yellow-brown colour and its strange smell which, if you were unlucky enough ever to catch a whiff, was reminiscent of mustard plants or horseradish.

Mustard gas was made using sulphur dichloride treated with ethylene, although a range of other chemicals, including hydrochloric acid, could also be used. When a chemical shell exploded, spreading a fog of mustard gas, soldiers had not only to put on respirators to avoid breathing the gas, but also to cover every square inch of their skin. Mustard gas caused chemical burns that irritated the skin, making it itch and develop enormously painful blisters. If it got in your eyes it could cause blindness and if you inhaled it, it caused bleeding and blistering in your lungs. The gas could penetrate normal clothing and you could even be affected simply by coming into contact with someone who had been directly exposed. Because it lingered in the soil and on equipment, mustard gas could render entire areas uninhabitable for a time, a phenomenon for which the military coined the term 'area denial'. Saturating an area with mustard gas could force the enemy to abandon important fortifications for days, if not weeks.

The Allies had no mustard gas weapon of their own until they captured stocks of German gas shells and it then took almost a year before the British perfected their own version, which they used against German troops in September 1918.

Despite a ban on their use under the Geneva Protocol of 1925, chemical weapons were used by a number of nations in conflicts between the First and Second World Wars. Although they were used in only a few isolated incidents in Europe during the Second World War, the Japanese used mustard gas extensively in China and Asia. They did not deploy the gas or other chemical weapons against Westerners, who had the capability to retaliate in kind. Mustard gas and similar chemical weapons are known to have been used a number of times since the Second World War, leading to another treaty in 1993 reinforcing the international ban on chemical warfare.

One thing that emerged from studies of the effects that mustard gas had on victims was that they appeared to suffer from a reduction in their white blood cell count. Further research led to the development of mustine, the first chemotherapy drug used in treating cancer.

IN PRAISE OF STINGING NETTLES AND REPLACEMENT WINDOWS

(TITANIUM)

A bright, beaming smile, a round of golf and stinging nettles are three things that would not normally be closely associated. No golfer is going to be smiling too much, after all, if he spends his round foraging about in stinging nettles looking for a lost ball. Neither are these things really of much use militarily unless you train your troops to bite, club or sting the enemy to death. We might as well ask what tennis rackets, replacement windows and an 18th-century Cornish clergyman have in common.

In fact, it's the same thing as all of the above – titanium.

The Reverend William Gregor was the rector of St Andrew's Church in Creed, Cornwall, and had studied chemistry at Cambridge University. As an amateur geologist, Gregor embarked on a study of local minerals and in 1791 he discovered a new metal in samples from the Manaccan valley. He named his discovery manaccanite. A German chemist, Martin Heinrich Klaproth, coincidentally discovered the same metal the same year, naming it titanium after the ancient Greek gods, the Titans. Klaproth's name caught on – titanium, that is, not Klaproth – but Gregor is still credited with having first discovered the metal.

That was almost where the story of titanium ended. It was found to exist in various forms throughout nature, with stinging nettles having the highest concentration of titanium in the plant world, but it was troublesome stuff to extract from the ores in which it was present. Unlike other metals, it couldn't be produced by heating the ore until it was molten to separate the liquid metal. Isolating titanium required the use of complicated, costly chemical processes. It wasn't until 1910 that it was first produced as a pure metal in small quantities in a laboratory and experiments continued to find practical ways to produce titanium in bulk.

Of course, you have to have a very good reason to want to produce metal on an industrial scale – there has to be a market for it. As chemists and metallurgists continued to study titanium it became clear that Klaproth had chosen well in naming it. Like those original Greek gods, titanium had immense strength. It was incredibly light, yet was as strong as steel and it was highly resistant to corrosion, even in sea water. The initial market for titanium was with the military.

An experimental jet, the Douglas X-3, designed to investigate the effects of supersonic flight, was the first aircraft ever to use titanium in its construction. It was involved in a number of test flights for the United States Air Force and NACA , the National Advisory Committee for Aeronautics (forerunner to NASA) in the early 1950s. In 1953 the supersonic F-100 Super Sabre fighter jet made its first flight, again making extensive use of titanium in its construction. Titanium had proved its value in military aviation and for years would be known as the 'aerospace metal' but, as it became more widely available, it found a variety of other uses. The Soviet Union, for example, used titanium when designing their *Alfa* class nuclear submarines in the 1960s.

Like so many other materials, mass production techniques have made titanium cost-effective for use in all sorts of things that we see around us every day. The metal can be combined with other elements in alloys that are used to produce golf clubs or tennis racquets and because, like gold, it does not tarnish, it is used in making jewellery. Titanium is non-toxic, strong and hard-wearing, making it ideal for use in dental implants and artificial hip joints. In powder form it is pure white and is added to many plastics, most of which are actually a dull grey, to give them a long-lasting white finish, especially useful for things like window frames that endure constant exposure to sun, wind and rain. Titanium dioxide is even used in whiter-than-white toothpaste to give you that bright, beaming smile!

HOW TO STOP AN ELEPHANT CHARGING

(CALTROP)

How do you stop an elephant charging? In the old joke, the answer is 'Take away his credit card' and in the cartoon world, an elephant, the largest living land mammal, is terrified of a mouse, one of the smallest. To stop a charging elephant, all you have to do is dangle a mouse in front of it. Set the mouse loose to scamper around and you can send an entire herd of elephants into a complete panic. In the real world, you would need a bit more than the help of a friendly mouse if you had to face up to an angry elephant, yet something smaller than a mouse could do the job for you.

The caltrop, based on a thorn design created by Mother Nature as seed cases for certain types of plant, is a simple yet highly effective tool that was used for slowing or stopping charging cavalry, even those mounted on fearsome war elephants. You can make a caltrop by twisting two nails together in such a way that it forms four sharp points. The points need to stick out at angles so that, however the caltrop is thrown on the ground, three of the points will always form a tripod to support the fourth point that will always be jutting straight up into the air. Scattering a few hundred of these on the ground will drive horse, elephants, camels or poorly shod soldiers absolutely mad.

The Romans used caltrops against horses drawing chariots, they were used against cavalry during the First World War and by the American Office of Strategic Service (OSS) intelligence agency during the Second World War to sabotage or ambush vehicles by bursting their tyres. They are still used today by police forces and the military to puncture vehicle tyres, although there is far less call for them to take the mouse's place in turning back elephants.

MAKING THE WORLD GO ROUND

(WHEEL SPOKES)

Cycling has never been as popular as it is today. It is estimated that there are 130 million bikes sold around the world every year. People use bicycles to commute to work; to race on road and track; to go thrashing around on muddy woodland tracks; and even as part of their jobs. Police officers patrol city centres on bicycles and urban deliveries are still made by bike. The bicycle is not, of course, a military invention, although soldiers have been issued with bicycles ever since the first recognisably 'modern', chain-driven versions were introduced towards the end of the 19th century. Troops struggled ashore carrying bicycles during the D-Day landings in 1944, commanders hoping that squads with bicycles would be able to cover more ground a lot faster by road on bicycles – providing that they could first get them out of the water and across the sand dunes.

So if the bicycle is not a military invention, what's it doing in this book? The answer is, that it isn't. Not all of it, anyway. Only part of the bicycle counts as a military invention – the spokes. All bicycles have them (unless you count racing bikes with disc wheels, so let's not) and always have had them. Even the earliest, most primitive types that had wooden wheels shod with metal tyres had spokes.

Spokes have to be made and fitted in a precise manner that maintains the rim of the wheel as a perfect circle. In the days of wooden wheels you needed a wheelwright, also known as a wainwright (which is where the surname comes from) to make or mend the wheels on your carriage or on your bicycle. Building and mending wooden wheels was so much of a skill as to become an art form and doing the same for a modern spoked wheel is almost as complicated. That, of course, begs the questions 'Why bother with spokes? What's wrong with a solid wheel?'

Solid wooden cartwheels were heavy, turned slowly and provided a very rough ride for whoever was in the cart. Anything with solid wooden wheels was slow. Wheels with spokes, on the other hand, turn far more quickly, are much lighter and grip the ground far better. The very bottom section of a wheel with spokes, flattens ever so slightly as the spokes compress under the weight of the vehicle. This puts more of the wheel in contact with the ground, providing better traction. Wheels that grip the ground will always let you go faster than wheels that slip. The compression of the spokes also provided a basic form of suspension, absorbing some of the bumps and thumps as you trundled along.

The first wheels with spokes, therefore, were used by people who wanted to go faster, people whose survival depended on being speedy – charioteers. Horse-drawn chariots raced into battle carrying a driver and an archer. They had only a light fairing or their shields behind which to shelter but the spoked wheels of their chariot gave them the speed they needed to charge around the battlefield, the archer choosing targets and picking them off at will, with the spoked wheels helping to steady his aim. Those early chariots were used around 4,000 years ago by the Andronovo people in what is now Kazakhstan, making the spoke one of the most enduring military inventions of all time.

FIX BAYONETS!

(BAYONET)

One of the best weapons that an infantry soldier could have in his hands when a cavalryman came galloping towards him on a huge horse, swinging a large, razor-sharp sword, was a pike or spear. With a wooden pike, which might be anything from 10 feet (three metres) to more than 20 feet (six metres) long and tipped with metal, you could deal the cavalryman a nasty blow before he got close enough to lop off your head with his sabre. Of course, if you failed to bring down the horse or rider with your pike, you needed to be a really good runner, and even if you did put your pike to good use on one rider, you had to get your act together pretty quickly in order to deal with the next one.

During the late 16th and early 17th centuries, when muskets started to become more commonly used by infantry soldiers, pike bearers were used to protect the musketeers. The firearms were slow to load, so a musketeer might only be able to take one or two shots at an advancing enemy before they were upon him. The musket was no use at all for close-quarters fighting and the musketeer would have to abandon it in favour of a sword or other weapon.

Then someone hit upon the idea of combining the reach of the pike with the firepower of the musket. Once the musket's usefulness as a firearm was over, why not fix a spike on the end so that it could be used like a pike? A musket was around five feet (1.5 metres) long and adding a lengthy blade to the end of the barrel turned it into a formidable pike-like weapon.

The name given to the blade was the bayonet, said to have been named after the French town of Bayonne, once famous for producing knives and swords. The first bayonets simply plugged into the end of the musket barrel, preventing any further use as a firearm. By the end of the 17th century, the bayonet was being used with a ring attachment that fitted around the barrel, allowing continued use of the musket and that basic system, in various guises, continues to be used today.

The actual bayonet has appeared in many shapes and forms over the years, from a simple spike to a double-edged 'sword' knife that could be used as a sword when not attached to the musket barrel. The bayonet also became a multi-purpose tool, with some designed to double as entrenching tools for digging, and the modern British Army bayonet able to be used, in conjunction with its scabbard, as a wire cutter.

The introduction of the bayonet brought with it a new infantry tactic, the bayonet charge, where a body of men, having discharged their weapons, ran at the enemy, their bayonets fixed to their muskets and pointing at the foe. As a military tactic, this became increasingly less viable as rapid-fire weapons began to dominate the battlefield but, having run out of ammunition and options, modern-day British soldiers fighting in Afghanistan have still been known to fix bayonets and charge at the enemy. In 2009, Lieutenant James Adamson was awarded the Military Cross for his bravery in continuing to engage the enemy when all he had left to fight with was his fixed bayonet and Lance Corporal Sean Jones also received the Military Cross for a similar action in 2011. Everyone knows that Wilbur and Orville Wright were the first men

SAILORS
IN THE AIR

(AIRCRAFT CARRIER)

to take to the skies in an aeroplane that flew under its own power. Their Wright Flyer made its first successful flight in December 1903, but they were almost pipped at the post by a flying machine sponsored by the US Army, with the US Navy taking particular interest. The Langley Aerodrome was launched from the deck of a converted houseboat in October 1903 but, while unmanned scale models had made successful flights under their own power, the pilot of the full-size version found more use for his swimming trunks than his flying helmet. The full-sized Aerodrome proved impossible to control, but that early attempt at powered flight came from what has to be considered the first aircraft carrier!

American military interest in powered flight was dampened almost as much as the Aerodrome pilot, making it difficult for the Wright brothers to be taken seriously by the army or the navy when they claimed to have solved the problems of controlling an aircraft in flight. They hardly helped their own case by insisting on keeping the details of their design secret, refusing even to mount a demonstration for the military before a purchase contract was signed.

Military men came round to the idea of powered flight again before long, of course, always trying to find better ways of putting 'eyes in the sky', not only to be able to see the enemy, but especially for artillery observation. Balloons had long been used to spot where artillery rounds were falling and to direct the guns to adjust their aim, but balloons needed to be tethered, couldn't be used in windy weather and were large, tempting targets for the enemy. It took a brave artillery observer to stand in a wicker basket below a balloon filled with highly flammable hydrogen as the enemy took pot shots at him.

What was needed was an aircraft that could move around of its own accord, making it more able to view enemy positions and less easy a target. Nowhere was such a mobile aerial observation platform needed more than at sea, where the extreme range of naval guns made spotting where the rounds fell most difficult. By 1910, when aeroplanes still looked like frail giant kites made of string and paper, a machine was successfully launched from a ramp on the bow of the cruiser USS *Birmingham*. The civilian pilot, Eugene Ely, got his boots wet as his aircraft plunged towards the water, but it pulled up, skimming the surface, for him to fly on and land on a nearby beach. In January the following year, Ely landed a plane on the deck of a ship for the first time, touching down on a specially built platform on the USS *Pennsylvania*.

The First World War saw incredible advances in aviation, with the aeroplane being transformed from an observation platform into an offensive weapon, with purpose-built fighters and bombers evolving. Sea planes, which had large pontoon floats rather than landing wheels, were carried by warships and lowered into the water to be launched, carrying out the first attacks by aircraft 'launched' from a ship, but the concept of the aircraft carrier took a giant leap forward with HMS *Argus*. At the outbreak of the war in 1914, the ship was being built as an ocean liner but was requisitioned by the Royal Navy and converted so that her entire upper deck was one long, flat runway. Aircraft could be raised on lifting platforms from a hangar below for launching. The *Argus* was the first real aircraft carrier and pioneered many of the techniques and systems that would later become standard, as well as solving some of the stability problems inherent in such an ungainly ship design. She entered service in 1918, too late to see action in the First World War, although she played an important role in ferrying aircraft to Malta during the Second World War.

Between the wars, aircraft carriers designed and built solely to launch and retrieve combat planes were one of the real growth areas in military development and they were to play a major part in every theatre of naval operations during the Second World War. The carriers enabled an attack force of aircraft to strike other naval vessels or shore targets well beyond the range of planes flying from a country's forward air bases. Without carriers, the Japanese would not have been able to launch the surprise attack on Pearl Harbor in December 1941 that brought America into the war.

Five months later, aircraft carriers made history when the Battle of the Coral Sea raged for four days between 4 and 8 May 1942, becoming the first naval engagement where the warring factions were able to savage each other without either side's most important ships having direct sight of the enemy or being able to fire on each other. It was the first battle fought between aircraft carriers and the killing blows were dealt by the bombs and torpedoes of the carrier-borne aircraft rather than the big guns of battleships.

Today, despite long-range aircraft and missiles, as well as their own vulnerability to attack, having threatened to make the carrier obsolete, the changing nature of conflict around the world has ensured that the aircraft carrier still has a valuable role to play. So-called 'gunboat diplomacy', where the threat from the immense firepower of a modern carrier is used as the 'muscle' behind political negotiations, often relies on the versatility of carriers like the United States' *Nimitz* class ships. Powered by two nuclear reactors, at 1,092 feet long (333 metres), they are twice the size of HMS *Argus* and can carry up to 130 jet fighter bombers. That constitutes an air force capable of delivering destructive power, including nuclear weapons, many times greater than all of the bombs dropped by the Allies during the whole of the Second World War.

The supersonic jets on a *Nimitz* class carrier would be like spacecraft from another world to Orville and Wilbur Wright, but they would surely be pleased to see the military finally taking aeroplanes seriously!

THE SARIN MENACE

(SARIN)

RAF Leading Aircraftman Ronald Maddison was looking forward to a few days off, time he would spend with his girlfriend, Mary Pyle. His three-day pass was part of his reward for taking part in a routine experiment at Porton Down, the military research establishment in Wiltshire, England. The other part of the reward was an extra 15 shillings in his pay packet.

In today's decimal currency, 15 shillings amounts to just 75 pence, or £0.75. When Maddison was offered the bonus in May 1953, however, 15 shillings represented half a day's pay for the average British worker. He was going to use the money to buy an engagement ring for Mary. He never got the chance.

Maddison was one of six test subjects who had volunteered to take part in experiments aimed at calculating the precise lethal dose of the sarin nerve agent. He was wearing a respirator and overalls, the same as the others sitting in the test chamber alongside him, and had two layers of uniform cloth taped to his arm. The sarin was applied to the cloth as 20 drops of 10 mg. Within about 20 minutes, Maddison started to feel unwell, sweating profusely, the cloth patch was immediately removed from his arm and he was helped from the chamber. When he then collapsed, antidotes were administered and an ambulance was called to rush him to hospital, but he died within 45 minutes. Just a few small drops of sarin, not applied directly to the skin or inhaled, had caused the death of a fit, 20-year-old serviceman in less than an hour.

Sarin had been developed in Germany in 1938 by scientists working on pesticides and was named after the men who developed it – Schrader, Ambros, Rüdiger and Van der Linde. The German Army put the chemical into production for use as a weapon, to be packed into artillery shells. It was never used as a weapon during the war, although versions of sarin have been used in terrorist attacks. In 1995 members of the Japanese Aum Shinrikyo caused the deaths of 13 people when they released sarin on the Tokyo Metro.

Exposure to sarin, which affects the central nervous system, initially induces flu-like symptoms – a runny nose, tightness in the chest and difficulty in breathing normally. It also affects the eyesight and before long brings on nausea and drooling. Loss of other bodily functions then follows, along with muscle spasms, loss of consciousness and death.

The use or stockpiling of sarin was banned by the United Nations Chemical Weapons Convention in 1993.

WELCOME TO RPG AVENUE ...

(ROCKET-PROPELLED GRENADE)

From the moment that the tank first appeared on the battlefield, weapons designers began dreaming up ways of destroying them. During the First World War, American physicist Dr Robert Goddard came up with the idea of a rocket-propelled missile fired from a tube. The device was successfully tested in 1918 just as the war ended. Too late to be put to use on the battlefield, the project was forgotten about.

During the Second World War, the need for such a weapon led to the idea being rediscovered and in 1942 the tube-launched, rocket-propelled, anti-tank weapon known as the 'Bazooka', entered service with the US Army. The Bazooka was so named because the tube looked like a home-made musical instrument called a 'bazooka' that was a little like a trombone. The Germans called it *Panzerschreck*, 'tank terror', and had their own version called the *Panzerfaust*, 'tank fist', that entered service around the same time.

It was from one of the German Panzerfaust designs that the Soviet Union developed their RPG2 in 1947, the forerunner of the world's most famous anti-tank weapon, the RPG7. RPG, taken simply to stand for Rocket-Propelled Grenade, is actually

Ruchnoy Protivotankovyy Granatomyot, 'hand-held anti-tank grenade launcher' but the difference is not likely to matter much to you if one is being pointed in your direction. Soviet troops were first equipped with the RPG7 more than 50 years ago and since then in excess of nine million have been produced for official use by more than 80 countries around the world and unofficial use by every terrorist group or criminal organisation that can get their hands on them.

Lightweight and easy to use, the RPG7 is instantly recognisable as a tube launcher with pistol grip and a rocket warhead protruding from the front of the tube. The launch tube rests on the user's shoulder and a variety of different sighting systems are used for aiming it. When the trigger is pulled an explosive charge shoots the rocket out of the tube. The missile's rocket motor fires after 10 metres of flight and fins pop out to stabilise it. Although the grenade can fly for more than 1 kilometre, it is most accurate over shorter ranges.

The weapon was used by the IRA during the troubles in Northern Ireland – one street in West Belfast became known as 'RPG Avenue' because the security forces came under attack there so often. The weapon was also used to shoot down two American Black Hawk helicopters in Mogadishu, Somalia, in 1993.

THE MIRACLE FIBRE

(KEVLAR)

Stephanie Kwolek always wanted to be a fashion designer. Born in Pennsylvania in 1923 to Polish immigrant parents, she spent her childhood sketching glamorous outfits and sewing garments. At school, however, she excelled at science and mathematics, leading her to choose an entirely different career path. Studying chemistry at university, she decided to take a job for a while after graduation in order to pull together enough money to carry on studying, hoping for a career in medicine.

The job Kwolek was offered in 1946 was at the DuPont chemical company, where she was to remain for the rest of her career. In 1964, Kwolek was part of a team investigating new polymers that might be used to create a lightweight nylon-like fibre that could replace steel reinforcement in car tyres making them lighter and helping to reduce fuel consumption. Her task was to produce chemical polymers that could be melted into a liquid and spun in a special machine called a spinneret that formed the liquid into fibres. One day Kwolek ended up with a perplexing liquid in her test tube that didn't look at all like she had expected it to. She decided to ask her colleague to spin it anyway and ended up with fibres that seemed really strong. When she baked them, they became even stronger and stiffer

Kwolek and her colleagues quickly realised that they were on to something and it became a team effort to develop what was known as poly-paraphenylene terephthalamide into a marketable commodity. Choosing the far snappier 'Kevlar' name probably helped. By the early 1970s, DuPont was ready to market its new textile, and there was immediate interest in using it as had originally been intended, in car tyres. There was far greater interest, however, from military sources.

The US Army began testing Kevlar as soon as it became available and the most obvious initial application for the new material was in body armour. So-called 'bullet-proof' vests had previously been made from cloth or canvas with steel plates inserted to protect the wearer. They were so heavy that they restricted movement, and being able to move quickly is a soldier's best defence against getting shot. Even when they were issued with them, soldiers often conveniently 'forgot' to wear them. A vest made from Kevlar, on the other hand, was almost as light as any other item of clothing, yet it was five times stronger than steel.

Before long, Kevlar was being used not only to make body armour, but to make combat helmets and to make armour for personnel carriers and even armour plating for tanks and aircraft carriers. On its own, Kevlar is capable of stopping low-velocity gunshots from hand guns and when combined with lightweight ceramic armour plates it can protect against more powerful weapons.

It did still make it into car tyres, as well as bicycle tyres, car bodywork, canoes, brake pads, racing-car fuel tanks, mobile phones and thousands of other products. The super-material does, however, have a Superman-style weakness. Its Kryptonite is sunlight. Kevlar can lose its strength when exposed to ultra-violet light and needs to be kept covered up or otherwise protected when used outdoors.

Somehow it seems a little unlikely that Stephanie Kwolek could ever have used fabric in quite so many extraordinary ways as a fashion designer.

SEEING IN THE DARK

(NIGHT VISION)

Being able to see in the dark has clear military advantages, but it has also proved to be a real boon to firefighters and other rescue workers operating at night. The main use for most night-vision systems is still, however, military and in some countries night-vision devices are restricted to military use, and banned from sale to the general public.

The first practical night-vision device was developed for the German Army during the Second World War and was called the *Nacht Jager*, or 'Night Hunter'. It was fitted to German tanks and involved the use of an infra-red searchlight which 'illuminated' targets using invisible infra-red light that was also, at the time, known as 'black light'. A powerful lens focused the light in an image intensifier tube, a vacuum tube similar to those used in the first televisions.

The system was also developed for use by infantry soldiers, allowing snipers to operate at night, although they had to carry a spotlight with an infra-red filter, the image converter and an array of batteries to power it all that more than doubled the weight of their kit. The German sniper version was known as *Vampir*, for obvious reasons, but the same technology was being developed almost simultaneously in America. In the latter stages of the war GIs were able to use 'sniperscope' night sights that they nicknamed 'snooperscopes'.

The great disadvantage of the infra-red systems was that they made everything, including the user, visible to anyone else using an infra-red night sight. As the technology developed, it was no longer necessary to illuminate targets by generating infra-red light, the systems relying on ambient light instead. At first, when the new generation of night-vision devices could magnify available light a thousand times, they really needed a little moonlight to operate effectively. The most modern versions can magnify ambient light by more than 50,000 times, producing a clear image using only starlight.

ICON OF
A GENERATION

(M16 RIFLE)

Some weapons can easily be said to be iconic designs and the M16 fits firmly into that category. Millions of photographs and countless hours of film footage have been taken of United States military personnel armed with the M16. Even someone who doesn't know its name would have no trouble identifying the M16 as the rifle carried by US troops in Vietnam, so distinctive is its design.

The M16 and a whole host of variants and descendants started out life as an ArmaLite design in the 1950s. ArmaLite was a private American manufacturer that set out to use the very latest advances in technology and engineering to produce thoroughly modern military rifles. ArmaLite aimed to make their rifles lightweight and innovative. In 1956 their AR-10 model was entered in a competition to find a replacement for the ageing Garand M1 rifle that had been in service since 1936. Unlike traditional rifles, the AR-10 had a straight-line, barrel-stock design. Other rifles had tended to have the stock, which nestles in the shooter's shoulder, lower than the barrel so that the sights, sitting on the barrel, were more in line with the shooter's eye to make it easier to aim the weapon. The AR-10's straight-line design had the rear sight on top of the fixed carrying handle and the front sight raised on a triangular support further along the barrel, making aiming the weapon quite simple, but unorthodox. The straight-line design allowed for a pistol grip below the stock, with both made out of fibreglass – far lighter than the traditional wood.

The AR-10 lost out to the heavier M14, which had traditional wooden bodywork and was essentially a development of the old M1. That was not, however, the end of the story.

Analysis of battlefield reports showed that up to two-thirds of all infantrymen never actually fired their weapons in combat, unable to find targets for their single-shot rifles. Soldiers armed with automatic weapons, on the other hand, were more likely to fire speculative bursts, and when two opposing combat teams came face to face, the team that fired the most rounds, irrespective of marksmanship, generally came out on top. The conclusion was that infantry soldiers should be armed with automatic rifles and in order that they could carry more ammunition, the bullets would be smaller.

The AR-10 was redesigned as a smaller weapon, the AR-15 and the Colt arms manufacturer, who had bought ArmaLite, began supplying the new lightweight assault rifle with its lighter ammunition to the US armed forces as the M16 in 1963. It had the same fixed carrying handle and sights as the AR-10 but was more compact to make it easier to use in urban or jungle warfare situations. It became the main service rifle for the US military, as well as being adopted by police forces and military units around the world including, at one time, Britain's elite SAS. Around 10 million M16 rifles have been produced in various guises, many of which are still in frontline service today.

The modern replacement for the M16 in US military service is the M4 carbine, a weapon that is a further development of the

M16 and can therefore trace its family roots back more than half a century through the ArmaLite range to the AR-10.

THE TOP SECRET FLYING CARPET FIASCO

(THE HOVERCRAFT)

Take a break, make yourself a cup of coffee, feed the cat and invent a new form of transport, but for goodness sake don't tell anyone because it's top secret. That's not exactly how it happened, but Sir Christopher Cockerell drew the inspiration for his ground-breaking new vehicle by watching the way that high-pressure air behaved when it was fed into the space between two tin cans – one small enough to fit inside the other. One was a cat food tin and the other a coffee can. The air flow was provided by a vacuum cleaner, not that he was combining his break with a spot of tidying up.

Cockerell, you see, was conducting his research on a shoestring budget. Having watched the way that speedboats rise out of the water and ride on air squeezed between their hull and the surface of the water, he had developed a theory about how a curtain of air around the outside of a boat could be used to contain an air cushion and sustain the effect. Speedboats lose the effect when they have to slow to turn as they need to have the hull in the water to make dramatic changes of direction. Cockerell envisaged a vessel floating on air like a flying carpet, experiencing no drag from having contact with the water which would allow it to go much faster and, more importantly, leave the water to travel over land as well.

Clearly this could have a huge impact on the way that troops came ashore during an amphibious operation. A landing craft packed with soldiers and equipment plodding towards the beach is a very easy target for gunners on the shore and when it hits the beach to disgorge its passengers, they also become sitting ducks. However, a boat that flies across the waves at high speed and then continues right across the sand to land troops where they have some cover could make an amphibious assault go far more smoothly.

Cockerell had reached the stage of building a working model and filing patents for his 'hovercraft' by 1955 but ran into problems when demonstrating it to military chiefs. British government ministers were terribly keen on the little machine that could drift across the linoleum on their office floors, float across the rug and back onto the lino. They classified it as top secret and brought in the military. The Royal Navy was impressed but saw the machine as an aircraft because it didn't go in the water. RAF bosses, on the other hand, saw it as a boat that occasionally took off, and the army didn't really know what to make of it.

With the military showing little interest in committing any portion of their budgets to developing the hovercraft, it was no longer deemed top secret and Cockerell was able to secure government backing to pursue the project with the British aero- and marine-engineering company Saunders-Roe. By 1959 they had built a prototype called the SR-N1 and they went on to develop bigger and better hovercraft, while also licensing the manufacture of hovercraft to other companies around the world. Sir Christopher Cockerell was knighted for his services to engineering in 1969.

Hovercraft are still used today in civilian and military capacities (including by Britain's Royal Marines), as landing craft, car ferries, gunboats, rescue vehicles and pleasure craft, but without the coffee can, the cat food tin and the vacuum cleaner the strange flying carpet idea would simply have floundered.

THE POWER
OF APOLLO
IN YOUR LAP

(THE PC)

Imagine having all the knowledge of Apollo – the god of medicine, music, poetry, sun, light and knowledge – in your lap, in your pocket or on a USB memory stick dangling from your key ring. Unfortunately, you can't have the power of a god in your key ring, but you can have the power of an Apollo space craft, or at least its computing power.

We all take computers very much for granted in the modern world. We use them every day in our cars, in our TV sets, in our MP3 players, in iPods, iPads, mobile phones and any number of other gadgets that clamour for our attention. The PC – Personal Computer, not Police Constable – is vital to most of us at work and in the PC we really do have the power of Apollo many times over.

Computers have been around for a lot longer than you might think. Charles Babbage designed the first programmable mechanical computer as long ago as 1837. It was a number cruncher that would have been able to manipulate numbers of up to 25 digits if he had ever actually finished building it properly. There were various other attempts to produce computers over the course of the next hundred years or so, but real progress came at the dawn of the electronic age during the

THE POWER OF APOLLO IN YOUR LAP

Second World War. The British Colossus computer was used to help break German wartime codes and was in use by 1943. It was noisy, it used 1500 valves (glass vacuum tubes used to control electric currents) in its CPU (Central Processing Unit) and it filled an entire room.

Advances in electronics meant that transistors began to replace valves in the 1950s and in the 1960s transistors were miniaturised to be embedded in integrated circuits. This is where the power of Apollo comes in. The Apollo Guidance Computer (AGC) was the first computer to use integrated circuits allowing the computer to make the navigational calculations that, 20 years earlier, could only have been done using a machine the size of an elephant. There was no room for an elephant anywhere aboard the Apollo spacecraft. Making electronic components more compact was a necessity for the space programme but a boon to everyone else when the technology spilled out into the marketplace.

Once engineers began to understand how to miniaturise electronic circuitry, breathtaking breakthroughs in computer technology came thick and fast. By the time IBM introduced their Personal Computer in 1981, it was able to perform complex functions, including word processing – or typing as it used to be known – and it had at least eight times the memory of the AGC. Within another ten years the first palmtop computers or personal organisers combined with mobile phones began to appear and by the beginning of the 21st century, you could stick Colossus in your pocket.

The average laptop or PC nowadays is immensely more powerful than the AGC that took Apollo to the Moon, even if you do only use it to store photographs of your cat and Skype friends who live only round the corner, but without the integrated circuits first used in the NASA space programme you could only manage that if you were the God of Knowledge.

LET'S GO
FLY A KITE ...

(KITES)

When George Banks took his children, Jane and Michael, to the park to fly their kite, watched by Mary Poppins who was about to take her leave of them, it was a wonderful, happy ending to a much-loved film and the cue for a great song. There was nothing warlike about it. Mr Banks had no military ambitions whatsoever. He was simply determined to enjoy some time with his children, playing with their toy kite ... but string-and-paper kites have not always been mere playthings.

Like so many good ideas, the kite originates in China. We can only speculate about how the first kite was devised but, because the Chinese had lightweight-yet-strong silk fabric (and later paper), lightweight-yet-strong silk thread to be used as string, and lightweight-yet-strong bamboo to create a frame, it would take only some casual ingenuity to watch a piece of cloth blowing in the wind and dream up a way to keep it under control. Once you've mastered flying your little kite, it can clearly be seen for miles around and that, around 3,000 years ago, was of great interest to the army.

Flags were used for signalling but had the disadvantage of only being visible if you were in direct line of sight with the flag bearer. A kite, however, could be seen from miles away, even, if it was flown high enough, from the other side of a hill. Kites had an obvious use for signalling, but it didn't take long before it was realised that, if someone on the other side of a hill could see a kite, then someone strapped to a kite could see whoever was on the other side of the hill.

Larger kites were constructed, capable of taking a man aloft to spy on the enemy and there is even a story about a battle fought during the Chu-Han War around 203 BC when kites were used for psychological warfare. The battlefield was fog-bound but, even though they couldn't see farther than the end of their spears, the armies could still hear each other. General Zhang Liang could hear the enemy soldiers singing songs about home and he struck upon the idea of flying large kites (you have to accept that there was enough wind amongst all that fog) over the enemy lines. Strapped to the kites were small children with flutes, who played tunes from the enemy homeland. When the soldiers heard this strange music coming, apparently, from nowhere, they were so frightened and homesick that they packed up their kit and left the battlefield.

Mary Poppins' 'Let's Go Fly a Kite' was almost certainly not one of the songs played by the intrepid kite kids, but the kite gave army commanders the idea of observation from on high and from kites to balloons, aircraft and satellites, they've been doing it ever since.

HOW TO SWIM
LIKE A FISH

(AQUA LUNG)

No one who remembers watching *The Undersea World of Jacques Cousteau* will ever forget the infectious nature of the French diving expert's love of the sea, even though it was a romance that Cousteau never intended.

The young Jacques Cousteau entered the French Naval Academy at the age of 20 and graduated as a gunnery officer, intent on pursuing a career as pilot with the French Navy. It wasn't until a car accident left him unfit for flying duties that Cousteau turned his attention to diving. Cousteau began experimenting with diving equipment in 1936 and in 1939 he first used the Self-Contained Underwater Breathing Apparatus (SCUBA), invented by fellow French naval officer Yves le Prieur.

The scuba gear involved the use of a tank of compressed air and a face mask that allowed the diver to breathe underwater unencumbered by tubes or other devices feeding air from the surface. Cousteau, who worked with French naval intelligence, could clearly see how useful scuba technology would be to navy divers involved in espionage, but was unhappy with the length of time he could spend underwater. By adding a demand regulator to the air supply, he was able to improve the system, allowing the diver to draw air as he breathed rather than being supplied with a constant stream. This made the air in his cylinder last longer.

Jacques Cousteau received several bravery awards for his work during the Second World War in operations against Italian espionage agents in France and also began making underwater films that led to his subsequent career as a TV naturalist and marine conservationist. He also worked with engineer Emile Gagnan, who had invented the regulator that Cousteau adapted for use with the scuba system, to perfect its use and make it ready for exploitation in a wider market after the war. The same basic system is still used by divers today.

Had he followed his original plan and become a navy pilot, the millions of people around the world who enjoy scuba diving on reefs and shipwrecks, as well as professional divers who work underwater on construction and engineering projects every day, not to mention fans of his TV series, might never have heard of Jacques Cousteau.

DREAMS AND NIGHTMARES

(AIRSHIPS)

Francesco Lana de Terzi is often referred to as the 'Father of Aeronautics', even though the 17th-century Jesuit mathematician never actually flew himself. He did, however, dream of building the world's first airship and his design, in theory, *should* have been able to fly.

Francesco's airship looked very much like a real ship, with a hull and a sail. The big difference was that as well as the mast for the sail, it had four other masts, each supporting a giant copper ball. His idea was that the balls would be made of incredibly thin copper and that, once all of the air was pumped out, leaving a vacuum inside, the vacuum balls would be lighter than air. They would then provide the buoyancy required to lift the airship off the ground. Unfortunately, it was not possible to produce copper thin enough in 1663 to produce the vacuum balls. Even if it had been, pressure from the outside air would have collapsed the balls, and copper thick enough to withstand the pressure would have been too heavy.

In any case, Francesco realised that his airship maybe wasn't such a good idea. He came to the conclusion that God would never allow anyone to build such an infernal machine because it would make every city on earth vulnerable to air raids when 'fireballs and bombs could be hurled from a great height'. Francesco may have been wrong about his copper balls, but he was right about the air raids.

Inscribed under the Arc de Triomphe in Paris, in recognition of his service to the French Army, is the name of Jean Baptiste Marie Charles Meusnier de la Place, which may explain why the Arc is so big. Meusnier designed an airship in 1783 that consisted of a long balloon containing a number of smaller, gas-filled balloons to provide lift for a boat-like cabin where a number of men would hand-crank large propellers to push the aircraft along. Meusnier had, in effect, invented the airship.

A number of different types of airship were developed over the years. Non-rigid airships rely on air pressure and the lifting gas to maintain the shape of the outer skin. Rigid airships have a structure of ribs to keep the outer shape with gas cells or balloons inside providing the buoyancy. Other types have some structural architecture to maintain their shape or even an outer skin made entirely of metal, something that would make old Francesco green with envy. Often referred to as 'dirigibles' from the French word *diriger*, meaning 'to direct', the fact that you can steer airships is what makes them different from balloons. All sorts of balloons have been used for military purposes, mainly observation in the case of balloons, but being able to steer an airship in powered flight, as opposed to a balloon being at the mercy of the wind, makes it of far greater use militarily.

The most famous dirigibles were the rigid Zeppelin airships, conceived by German Army officer Count von Zeppelin. The first Zeppelin was launched in 1900 and, despite some initial setbacks, the design became so successful that it turned the Zeppelin name into a generic term for airships. Zeppelins conducted the first air raids on British cities during the First World War, making

Francesco's fears a reality 252 years on, although the airships were not hugely successful in this role. As aeroplane technology improved, fighter pilots found the airships easy prey.

Airships still had the capacity to carry greater loads over longer distances than aeroplanes and between the First and Second World Wars they established trans-Atlantic flights, carrying passengers in accommodation with the comfort of an ocean liner, long before aeroplanes were capable of such feats. When the Empire State Building was completed in 1931 it had an airship mooring mast at the top, although the blustery winds blowing an airship around at 1,250 feet (381 metres) would have made disembarking a daunting prospect. High-profile disasters like the *Hindenburg* crash, when the giant airship burst into flames while coming into land in New Jersey after a trans-Atlantic flight from Germany, brought the airship era to an end. The *Hindenburg* burned ferociously because it used highly flammable hydrogen as its lifting gas. Had they used the inert gas, helium, there would have been no fire from a gas leak. The passengers might have disembarked talking like Mickey Mouse, but the airship would not have burned. Unfortunately most of the world's limited supply of helium was controlled by the United States, who refused to sell it to Germany.

Although airships were used by the military during the Second World War, mainly for coastal submarine-spotting patrols, their use during the latter part of the 20th century was confined to civilian roles in advertising, sightseeing and as camera platforms. In the 21st century, however, there has been renewed interest and a great deal of military research conducted into using airships as long-range, heavy-lifting cargo vessels and as high-altitude communications stations.

Francesco would, no doubt, be proud that his dream came true, even if some of his nightmares came along with it.

THE TIN FISH

(TORPEDO)

Have you ever heard of the Austrian Navy? Why would Austria, a landlocked country, need a navy? Austria has not always, of course, been landlocked and those of you who are fans of *The Sound of Music* will recall that Baron von Trapp had been a submarine commander in the Austro-Hungarian Navy during the First World War. At that time, the Austro-Hungarian Empire included ports on the Adriatic such as Fiume, where one of the most devastating naval weapons of all time was developed – the torpedo.

The name 'torpedo' has been used for all sorts of explosive weapons over the years, including naval mines and booby traps, but in the latter part of the 19th century it was applied to a long, tubular device that carried an explosive charge of nitrocellulose, commonly known as guncotton. The tube, actually called *Minenschiff*, 'mine ship', had been developed for the Austrian Imperial Navy in 1866 (the Austro-Hungarian Navy came later) by English engineer Robert Whitehead, based on an idea that had been suggested to him by an Austrian naval officer.

With Austrian government backing, Whitehead set up a factory in Fiume, then the home port of the Austrian Imperial

Navy and now called Rijeka in Croatia, to manufacture the weapon. He refined his design into a device that was powered by a motor that ran on compressed air turning a propeller at the rear. The torpedo, as it became known, soon picked up the nickname 'tin fish'. It ran at a preset depth due to a clever system – a hydrostatic valve that was sensitive to water pressure and sensed the depth of the torpedo. The valve activated a pendulum which then adjusted the hydroplanes to keep the torpedo running at the required depth. The torpedo could travel at about 7 mph and hit a target over half a mile away.

Whitehead soon had orders for his torpedo from navies all over the world, including Britain's Royal Navy and in 1871 he struck a deal with the British government to begin producing torpedoes in the United Kingdom.

The original torpedo design underwent constant improvement, making it faster, larger and more accurate, and allowing it to carry a more powerful explosive. In 1878, a Turkish ship called the *Intibah* became the first to be sunk by torpedoes when she was attacked by a Russian torpedo boat during the Russo-Turkish War but it wasn't until the First World War that the torpedo really came into its own.

By 1914 torpedoes had been developed that could be launched by boats, dropped from aircraft and, crucially, fired by submarines. Submarine torpedo attacks on supply convoys heading for Britain during the First World War helped to define the strategic use of the submarine and by the Second World War the torpedo had become every sailor's worst nightmare. Torpedoes that would detonate not on contact, as they had done previously, but beneath the ship, or close to the ship, when the steel of the target ship's hull activated a magnetic trigger were in widespread use. They also had electric motors giving them longer range and they could even home in on the sound of their target, making avoiding a torpedo once it had been fired at you almost impossible.

Modern torpedoes, of course, use technology that makes them more deadly than ever before. Torpedoes can now be launched by missiles or even lurk at the bottom of the ocean until an enemy ship comes within range and launch themselves.

The torpedo is still very much a part of modern naval warfare but Baron von Trapp's Austrian Imperial Navy, for which the first torpedoes were made, is no more. From the mid-1950s the navy was reduced to just two patrol boats on the Danube, which were actually operated by a naval unit of the army, and they were withdrawn from service in 2006.

FOUNTAINS OF FLAME

(FLAME THROWER)

The flamethrower is one of the most hideous modern battlefield weapons ever devised, created for the German Army at the beginning of the 20th century as the *Flammenwerfer*, literally 'flamethrower'. Or was it? The flamethrower has a longer history than you might think.

In 1901, the weapon was conceived as a pressurized cylinder containing gas and flammable oil. When a lever was pulled on a hand-held nozzle, a bit like the nozzle on a fire hose, that was attached to the cylinder by a hose, it opened a valve that allowed the gas to push the oil out of the cylinder, along the hose to the nozzle where it was ignited by a naked flame as it shot out of the nozzle.

The Germans used their flamethrower in the trenches during the First World War, although it was an awkward, heavy piece of kit for anyone to try to use in the confined space of a trench and it actually required two men to operate it. One carried the gas propellant tank and the other carried the oil tank and nozzle arrangement. With a range of around 60 feet (18 metres) it could be fired from German trenches to land oil and flame in Allied trenches. If the flames didn't get them, the soldiers on the receiving end of the *Flammenwerfer* were shot as they abandoned their trench.

The same basic system was used in the Second World War, although by then it had been refined and reduced to become a single operator device, although in many cases the weapon was still deployed in two-man teams. The man who had to stand up and aim the flamethrower at the enemy, however, was a prime target for every enemy soldier capable of aiming a rifle. Flamethrower teams stood a very good chance of going up in a fireball, the victim of their own device.

A safer way of using a flamethrower, for all but those on the receiving end, was to mount the nozzle on an armoured vehicle. A tank carrying a flamethrower could use bigger pressure cylinders, giving it greater range. A tank could set a whole hillside alight to flush out enemy troops. They were used by both sides during the war, but saw extensive use against the heavily fortified Japanese bunker positions during the Pacific campaign.

Flamethrowers used by infantrymen, tanks and naval vessels saw service during the wars in Korea and Vietnam but they are not widely employed by the military in the 21st century, having been banned by international treaty. Banned after a century of use in modern warfare? Not quite. Banned after two millennia of use might be more accurate. The ancient Greeks used 'Greek fire', a mixture of oil, sulphur and other flammable agents, that was pumped by hand to spray out through tubes, setting fire to enemy ships or scorching enemy troops.

The horror of the flamethrower has been around for longer than we might realise and the only reason that modern armies no longer use it is because they have far more effective weapons at their disposal.

LIFE-SAVING FUNGUS

(PENICILLIN)

Necessity is the mother of invention, so the saying goes, with Greek philosopher Plato most often being given credit for the ancient soundbite, but it was never more apt than during the Second World War when front-line soldiers, wounded in battle, needed the best medical treatment available.

One of the biggest battlefield killers was secondary infection. Under combat conditions, a relatively minor wound that was not life-threatening could quickly turn into a killer if it became infected. In ancient Egypt, China and South America, medical men knew that certain moulds could be used to treat infected wounds, although they didn't understand how or why they worked. It wasn't until the 19th century that scientists like Louis Pasteur in Europe began to appreciate how germs caused diseases.

In 1928, Scottish scientist Alexander Fleming, working in a laboratory in St Mary's Hospital in London, noticed that a mould had grown on a petri dish that he had accidentally left uncovered. The petri dish contained a bacterial culture that he was working with but, around the edge of the blue-green mould that had developed, he could see that the bacteria were being killed off. He examined the blue-green growth and found it to be a *Penicillum* mould, a natural fungus group with more than 300 species. Fleming used the term 'penicillin' to describe the culture that he developed from the mould, and began researching its properties. He believed that it would make a powerful antiseptic and later began experimenting in using it to combat harmful bacteria.

By the time the Second World War began in 1939, a team from Oxford University was working on ways to use penicillin to treat patients with infections, but there wasn't enough penicillin available in drug form for extensive testing. As the war grew more intense, resources in Britain became scarce and a small quantity of penicillin was taken to America where work began on finding a way to produce the drug in quantity. Clinical tests ultimately showed that penicillin was the most effective antibiotic yet discovered. It was rushed into production on an industrial scale, bringing the cost of manufacture tumbling down. The small sample that had been brought to America was so valuable that no real price could be put on it, but by 1943 the cost of penicillin was down to $20 per dose. Within two years that price would drop to less than a dollar.

When the Allies invaded Europe on D-Day in June 1944, there were 2.3 million doses of penicillin available to treat casualties and by the end of the war in 1945 more than 600 billion doses were being produced every year. Today, it remains one of the most widely used antibiotics in the world.

The military may not have played a direct role in discovering penicillin, but their pressing need for the drug was what galvanised the research and the clinical trials and what put it into volume production.

THE GUNSLINGER'S BEST FRIEND

(REVOLVER)

The image of the gunslinger standing facing his adversary in the dusty main street of a western town, poised and ready for a quick draw, his hand hovering over the six-gun in his holster, has been at the heart of so many Hollywood movies that it surely must have been a weekly occurrence in the 'Wild West'. Actually, like so many images produced by Hollywood – and we should all be thankful for the escapist fantasy entertainment of the movie industry – gunfights in the old west very rarely happened this way.

Most gunfights erupted at very close range, an argument exploding out of control when the gunslingers reached for their revolvers. At such close range, the revolvers they were reaching for, most likely the ubiquitous Colt 45, also known as the Peacemaker, were deadly. At a range of 50 yards (45 metres) it would take an expert marksman to hit anything, even with a calm and steady hand, which your gunslinger fighting for his life in a quick-draw contest was unlikely to have.

The Colt was not the first firearm to use a 'revolver' concept. The Chinese used revolving barrels on some of their early weapons and there were even a number of revolving flintlock designs but, while they gave an improved rate of fire when they worked properly, they were cumbersome and slow to reload. In the early 1830s Samuel Colt came up with the idea for a revolver pistol design using a cylinder that contained five chambers. Loading it involved dismantling the thing and filling each chamber with gunpowder and a lead ball, then placing a percussion cap on the end of each chamber using a special tool. The Colt Paterson looked like a modern revolver, but a gunslinger who wanted to be able to shoot fast and reload quickly had to wait for the invention of the metal cartridge and the introduction of the Smith & Wesson Model 1 in 1857. The Model 1 had a seven-shot revolving cylinder and was a single action weapon, meaning that the hammer had to be cocked for each shot, and in doing so the cylinder revolved to present the next round. When the trigger released the hammer, it struck the percussion cap in the end of the cartridge, detonating the powder and sending the bullet rocketing down the barrel.

The system was ultimately adopted by other manufacturers, once patents covering the Model 1 had expired or could otherwise be circumvented, and the Colt 45 was designed to meet a specification issued by the US Army in 1872. Like the Model 1, it was a single-action weapon, which is why you may see movie gunslingers firing the weapon with their right hand but 'palming' the hammer back with their left. The '45' in the name refers to the calibre of the ammunition it used, the bullets measuring 0.45 inches in diameter. Many Peacemakers were actually made to take a slightly smaller round in order to make the ammunition interchangeable with the Winchester '73 rifle, the gunslinger's second-best friend.

Double-action revolvers, where pulling the trigger cocks the hammer, revolves the cylinder and then releases the hammer, were developed in England in the 1850s and became increasingly popular in the late 19th century, although single-action pistols were often felt to be more accurate.

Because revolvers have relatively few moving parts they are simple to maintain, robust and reliable, which is what made them the preferred sidearm for police forces and armed services around the world through both the First and Second World Wars and well into the 20th century.

The gunslinger's Peacemaker continued to be made in special editions by Colt up to 2003 and in 2010, to commemorate the role it played in the state's 'Wild West' history, it was adopted by Arizona as its official state firearm.

MIND
YOUR STEP

(LAND MINE)

One of the biggest problems that ordinary people have to face when the fighting stops at the end of a long and messy war is tidying up afterwards. Areas that were fields, farmland and beaches before the war may have become military camps or outposts during the hostilities and returning them to their original purpose in order to let people start rebuilding their lives involves clearing them of dangerous munitions. Chief among those since the beginning of the 20th century has been land mines.

Throughout history, defending a village, a hill, a castle, a beach or just about anywhere else against attackers has always involved creating a few hidden surprises for your enemy. Whether that might be a concealed pit full of sharpened sticks, a trench filled with oil that can be set alight or trip-wire booby traps that bring trees, boulders or a hail of arrows raining down on the unwary, the hidden surprises needed to be suitably deadly to make anyone think twice about trying to sneak up on you again. The most deadly hidden device is the land mine.

As you might expect, the Chinese, experts at finding all sorts of different ways to make use of their gunpowder, were the first to come up with the idea of a land mine. By the 13th century, they even had trip-wires and pressure pads that could create sparks to light the fuses of bombs buried out of sight. Trespassers never knew what hit them.

Similar devices were tried in Europe as the use of black powder spread throughout the rest of the world, but the big drawback with burying gunpowder in the ground is keeping it dry. If it is allowed to become damp, it may well fail to ignite. By the end of the 19th century, other explosives were quickly becoming available and by the time the First World War began, the modern land mine had been developed.

Land mines come in many shapes and forms, but they all still work in much the same way as the ancient Chinese booby traps. A trip wire or pressure plate activates a detonator that sets off an explosive charge. Like a hand grenade, the casing of the mine becomes shrapnel that can maim or injure over a wide radius but whoever triggered the mine is caught in a blast that tears them apart, regardless of the shrapnel. Some mines are designed with a small charge that sends the main mine shooting into the air before it detonates, spreading the shrapnel over a wider area. Others are designed specifically to wound rather than kill whoever steps on them, perhaps just blowing off a foot or part of a leg. Military minds have calculated that, while a dead soldier will be left where he lies, a wounded man will need the attention of at least two others to get him to safety, removing three enemy personnel from the battlefield instead of only one.

Modern mines, in waterproof casings, can remain in the ground, live and dangerous, almost indefinitely. This has caused a massive problem in North Africa where millions of mines were laid during the Second World War. Those who laid the mines kept detailed plans to avoid stumbling into their own minefields, but after a battle, those who laid the mines and their detailed plans could not always be found. The Germans laid 80,000 mines in the Beurat area in Libya alone. Across Egypt's western desert it is estimated that there are around 19 million mines.

Civilian death and injury from land mines in conflict zones around the world led to the establishment of the Ottawa Treaty in 1997, banning the use or manufacture of anti-personnel mines, with 160 countries around the world signing the agreement. There are still up to 100 million anti-personnel mines stockpiled, ready for use, worldwide.

THE CHARMS OF BROWN BESS

(FLINTLOCK)

You might think that you know nothing about 18th-century firearms, but you're probably wrong. The flintlock was the weapon used by British infantry soldiers for more than a century, something that is unthinkable nowadays. Given the speed at which modern firearms have developed, it's hard to imagine that any soldier serving today will be using the same weapon as his counterpart in the 22nd century.

The flintlock first made its appearance at the beginning of the 17th century when the muskets in use by most European armies were essentially no different from the guns used by the Chinese 400 years previously. The musketeer had a smouldering rope, known as a match, dangling from his belt. Once he had rammed gunpowder, wadding and a lead ball down the barrel of his weapon, he charged the flash plate with a little gunpowder, aimed the weapon and then touched the flash plate with his match. The flash ignited the main charge in the barrel, firing the weapon.

To save him from fiddling around with his match in battle, the matchlock was developed, with the smouldering match on a spring-loaded hammer that could be pulled back, or 'cocked', and locked in position until the musketeer had taken aim, then released using a trigger. The hammer snapped forward, making contact with the flash plate.

The flintlock worked in much the same way as the matchlock, except that the hammer was fitted with a piece of flint instead of a smouldering rope. The flint scraped against a steel plate as it fell towards the flash plate, creating a spark. The great advantages were that the musketeer no longer needed to carry a lighted rope, which was never very safe when handling loose gunpowder, and he had both hands free to aim the weapon.

You might not have known any of that, but you almost certainly have heard or used phrases and sayings that date all the way back to the flintlock era. There were two positions in which the hammer could be locked. The first pulled the hammer away from the flash pan to allow the pan to be charged with gunpowder. This was the safety position as the hammer would not generate a spark if it was released and the weapon would not fire. It was known as 'half-cock' and to this day if someone is said to have 'gone off half-cocked' then they weren't ready for what they tried to do. 'A flash in the pan' is another phrase still used, as is 'lock, stock and barrel', both referring to the flintlock.

The flintlock, with its walnut woodwork, was nicknamed 'Brown Bess' by British soldiers who carried the weapon into battle from 1722 until 1838. It was fondly recalled by Rudyard Kipling in his 1911 poem 'Brown Bess' –

In the days of lace-ruffles, perukes, and brocade
Brown Bess was a partner whom none could despise,
An out-spoken, flinty-lipped, brazen-faced jade
With a habit of looking men straight in the eyes,
At Blenheim and Ramillies, fops would confess
They were pierced to the heart by the charms of Brown Bess.

BAD NEWS FOR BULLET COUNTERS

(AUTOMATIC PISTOL)

'That's a Smith & Wesson, and you've had your six,' were the last words that Professor Dent ever heard. James Bond then shot him. The scene from the film *Dr No*, where Bond has waited patiently in a corner of a bedroom, with the bed made to look like he is sleeping there, is a classic for movie bullet counters. It became almost traditional for a revolver to have six chambers in its cylinder, allowing six shots to be fired before reloading. Although some revolvers do have more than six chambers, the 'six-guns' of the Wild West were the basis for most movie bullet counters' calculations. If a movie cowboy fired more than six without reloading, there was a great 'tutting' and a shaking of heads.

The automatic pistol changed all of that. More correctly termed 'semi-automatic' because you have to pull the trigger to fire each shot rather than simply holding the trigger while the weapon fires repeatedly, these handguns began to appear at the end of the 19th century when manufacturers applied the principles Hiram Maxim had used in his machine gun. The firing pin on the hammer strikes the percussion cap on the cartridge, which detonates the charge, the expanding gases from the explosion sending the bullet hurtling down the barrel. In order to push the bullet forwards, of course, the gas must also push back on the pistol and this recoil is used to move the slider that sits over the length of the barrel. The slider is what makes automatics look squarer than revolvers. The spring-loaded slider flips the spent cartridge out of the chamber and cocks the weapon as it travels backwards, then moves forward again on its springs to chamber a fresh round. The rounds are fed upwards from a spring-loaded magazine inside the hand grip.

There is far more space inside the hand grip to store ammunition than the simple six chambers of a revolver and some automatics have magazines that can take 15 rounds, while others may have only five. This is what foxes the movie bullet counters when the screen shoot-outs involve anything other than revolvers.

James Bond's counting in the bedroom shoot-out would have been accurate if Dent had been using a revolver, but as he was holding an automatic he could started with anything from five to 15 rounds in the magazine. Even if it was a Smith & Wesson automatic, it would have held at least seven in a full clip. Dent might have had several shots left, Bond might never have left the bedroom and Sean Connery might never have gone on to make *From Russia with Love* ...

SKILLS THAT ARE OUT OF THIS WORLD

(THE SWORD)

In parks and playgrounds the world over it never takes too long to find children involved in sword fights, whether they are using twigs backed up with clashing steel sound effects, or whether they are wielding nothing but fresh air backed up with a vivid imagination. They are pirates, outlaws with Robin Hood in Sherwood Forest, Hobbits on a quest or Knights of the Round Table following King Arthur into battle. They might even be Jedi Masters – well, a light sabre has to count as a kind of sword, doesn't it?

A sword is a blade weapon designed for cutting or thrusting, slashing or stabbing, depending on the style of the blade. A sword is, in effect, a knife or dagger on a bigger scale, but knives of all sorts existed long before technology allowed for the creation of the sword. The problem that early sword makers faced five thousand years ago was that the metal they had to work with, copper, wasn't really up to the job. With copper, they could craft a sharp and useful dagger, but when they tried to make the blades longer – and having a longer blade gave the obvious advantage of having a longer reach in a knife fight – the copper blades tended to bend. Size wasn't everything if you had a longer blade that tended to be a bit droopy. Using tin along with the copper to create bronze helped a bit, but bronze still didn't have the tensile strength required to make a significantly longer blade.

When the world moved on from the Bronze Age into the Iron Age around 1200 BC, iron blades were the things to have. They could still become bent out of shape when used in battle, but iron was more easily available and less expensive than bronze, as well as being easy to work with. By the 8th century BC, entire armies could be armed with iron weapons, although Ancient Greek and later Roman swords were still only around two feet (60 cm) long. It wasn't until metalsmiths discovered that adding carbon to molten iron created a tougher metal, steel, that longer blades became practical. Steel began to be widely used around 650 BC.

The swordsmith's job was quite an art, involving heating a billet, a bar of metal, in a furnace until it was soft and then forming it using a hammer and an anvil, with a variety of other tools coming into play as the sword began to take shape. To achieve the required hardness, strength, flexibility and balance in the sword, different grades of steel could be heated and folded together to create the blade and it might have ridges running down the flat of the blade to give it extra strength. The blade was worked with the hammer to move the metal around, tapering the edges and keeping the weight distribution absolutely right in order that the finished sword would be perfectly balanced. After being allowed to cool, the blade would be sharpened and polished before the swordsmith passed it on to other craftsmen to finish the hilt and add decoration.

From the slim and highly flexible rapier suitable for the swift movements required in fencing to the heavy broadsword that a strong man wielded using both hands, the range of blades devised by swordsmiths in different parts of the world is enormous.

The Jedi's light sabre, on the other hand, requires the skills of another world entirely.

A PRESENT
FROM POLAND

(MINE DETECTOR)

The British government had a bit of a problem when it became necessary to reorganise the country's coastal defences in 1941. By then it was clear that Hitler had changed his mind about invading Britain as he had turned his attention to the eastern front, launching attacks against his former ally, the Soviet Union. The tricky thing was that 350,000 land mines had been laid on Britain's beaches and nobody was exactly sure where they all were.

Given the nature of the shifting sands on beaches and the fact that the coastal defences had been set up in such a hurry, clearing the mines, or re-laying the minefields, looked like becoming a real headache. The War Office issued a specification for a device that could locate buried mines.

A Polish Army officer, Lieutenant Jozef Stanislaw Kozacki, was then working in St Andrews in Scotland, where a number of Polish regiments were stationed. Kozacki was a signals officer and had left Poland with his unit in 1939 to fight on in France, and had then been evacuated from France to the UK. Before the war, he had been working on a device that could be used to locate dud shells on an artillery range, and he perfected the detector while in Scotland, submitting it to the British War Office as a mine detector.

Kozacki's detector involved the use of two coils. One was connected to an oscillator that sent a current through the coil. The other was connected to an amplifier and a telephone earpiece. The current being passed through the first coil creates an electro-magnetic field, basically sending a signal into the ground. When it hits something metal in the ground, the signal bounces back to be picked up by the second coil, which acts like an antennae. This generates an audible warning that lets the person operating the equipment know that the coils on the end of the pole he is holding are hovering above a mine.

The device worked so well that, British beaches aside, 500 of Kozacki's mine detectors were rushed to North Africa to help Allied troops pick their way through German minefields at El Alamein. More than 100,000 of the mine detectors were produced during the course of the war.

Kozacki never patented his mine detector, offering it as a gift to the British and receiving in return a letter of thanks from King George VI. Mine detectors based on Kozacki's design were used by the British Army for over 50 years and it formed the basis of the modern metal detector used not only by treasure hunters but also for security scans at airports.

THE GI'S CADILLAC

(M1 GARAND RIFLE)

When American GIs went into battle during the Second World War, they did so carrying a rifle that was, in many ways, far superior to those used either by their Allies or by their enemies. In a letter to General Campbell, who was in charge of ordnance for the US Army, General George Patton wrote, 'In my opinion, the M1 Rifle is the greatest battle implement ever devised.'

What made the M1 so special was the fact that it was semi-automatic. Unlike the Lee-Enfield used by the British Tommy or the Mauser Karabiner 98K that was issued to German infantryman, the M1 was a semi-automatic rifle, not a bolt action weapon.

A British or German infantryman had to lift, pull, push and lower a lever to operate the bolt on his rifle, ejecting a spent cartridge, re-cocking the weapon, chambering a fresh round and locking the mechanism. Then he could pull the trigger. A GI pulled the trigger, the rifle fired, the cartridge was ejected, the weapon was cocked and a new round chambered without him having to move a muscle. The semi-automatic features made the M1 the Cadillac of combat rifles.

To achieve this semi-automatic performance, the rifle used some of the expanding gas from the detonation of the round being fired – gas that would normally all be used to push the bullet out along the barrel – to push the firing mechanism back into the cocked position, accomplishing the other functions along the way. It gave the GI a far greater potential rate of fire than his counterparts but, because only one shot was fired at a time, he could still maintain a steady aim. The M1 was accurate up to a range of 500 yards (454 metres).

The British Army considered adopting the M1 as a replacement for the Lee-Enfield but rejected it because it was believed that the semi-automatic mechanism would be prone to jam in the muddy conditions of a battlefield, so chose instead to stick to what it regarded as the more robust Lee-Enfield. In fact, the M1 proved itself to be a highly reliable and was the standard-issue service rifle for US armed forces from its introduction in 1936 up to 1957. It was exported for use by more than two dozen other countries including France, Norway, Greece and, after the Second World War, West Germany and Japan.

Well in excess of six million M1 Garand rifles were produced along with a similar number of the shorter 'carbine' version produced for airborne or mobile troops who would have found the standard rifle awkward to handle in cramped conditions – a kind of mini-Cadillac!

ARMY FASHION CLASSIC

(THE DESERT BOOT)

While serving as an officer in the British Army during the Second World War, Nathan Clark came across an idea that was to create a footwear fashion classic. Clark was the 25-year-old great-grandson of James Clark, who had co-founded the famous Clark's shoe company in Somerset, England in 1820. While he was doing his military service, Nathan took a natural interest in what people were wearing on their feet.

That, you might think, would be army boots or, in the case of officers at that time, officers' army shoes. What Clark found when he was posted to Burma in 1941, however, was that the officers had adopted a different style. Many of the officers there had served in the desert campaigns in North Africa and, when they were off duty, they wore low-cut rough suede boots with crepe soles and only a couple of lace holes. Normally, most officers would never have worn suede shoes, suede not being the sort of thing a gentleman could be seen in – too caddish and uncouth. In the desert, however, the sand and heat took a heavy toll on English leather, making it almost impossible to keep shoes or boots looking smart. Normal shoes were also hot, heavy and uncomfortable. Some officers took their lead from their South African counterparts, who had lightweight suede boots.

They had copies of the boots made in Cairo and found that the crepe soles were comfortable, also giving a better grip on gravelly sand than the smooth leather soles they were used to. The lightweight boots were just as good in the heat of Burma, although as non-regulation footwear they could not be worn on duty.

Clark made patterns for the boots out of newspaper and posted them home to his brother, who was chairman of the family business, but little attention was paid to the design until he returned to Somerset at the end of the war. Working with pattern maker Bill Tuxhill, Clark created a prototype and showed the boots at the Chicago Shoe Fair in 1950. There was immediate interest in the boots that could be marketed as British colonial footwear with 'plantation rubber' soles and the desert boot became a huge fashion trend, immensely popular with young men – especially because they never needed to be polished!

DON'T SHOOT THE PROPELLER!

(SYCHRONIZATION GEARS)

Pilots doing battle in dogfights during the First World War, the Knights of the Air, twisting and turning to bring their guns to bear on their elusive targets are the stuff of legend. Had it not been for some engineering ingenuity, they would have remained precisely that – legends, myths or fairytales. They would have had no practical, effective way of shooting at each other in one-on-one aerial duels at all.

The first problem that would-be aces had to overcome was the fact that some people really didn't want them to be able to shoot at one another. Aircraft, frail yet expensive machines, were extremely useful for observation, keeping an eye on the enemy, and if they started taking pot shots at each other it might prove to be something of a distraction. In the early part of the war, many aircrew went along with this notion, even saluting each other when they passed close by their opposite numbers.

Before too long, however, pilots and their observers began taking matters into their own hands, quite literally, by blasting away with rifles or revolvers whenever they came within range of the enemy. It soon became clear that aircraft could play a more aggressive role in the conflict. In an aircraft capable of carrying a pilot and an observer, it was simple enough to provide the observer with a machine gun to attack other aircraft or to defend himself and his pilot. That worked well as long as he didn't shoot his own plane's tail off, as Dr Jones Snr managed to do when flying with his son as pilot in *Indiana Jones and the Last Crusade*. The extra weight of an observer, a machine gun and its ammunition, however, made aircraft with two occupants far slower than single seaters, and consequently somewhat easier to shoot down, either from the ground or from the air. In the air, the priority was to find ways of arming the pilots of single-seat aircraft.

Mounting machine guns on the wings of a biplane, or one of the early monoplanes, was not always possible because of the way the wings were braced and stressed using struts and cables. The wings of these aircraft were essentially kite-like structures that were not strong enough to take the weight of a heavy machine gun or the stress that it caused when it was fired. Some biplanes were fitted with a machine gun above the top wing, firing forwards, but this was difficult to aim properly and if he needed to reload or deal with a jamming problem, the pilot had to stand up to reach the gun. His cockpit, of course, was open to the elements.

The ideal solution was to mount the machine gun on the fuselage, fixed in position, firing forwards so that the pilot could bring it to bear by pointing the nose of his aircraft at the enemy and have it within easy reach. The only problem was that, with the majority of aircraft, the propeller was in the way. Obviously, all that was needed was to be able to fire through the spinning propeller without the bullets chewing the propeller blades to pieces. Both the Germans and the Allies came up with mechanical solutions that synchronised the gunfire with the rotation of the propeller so that the machine gun would not fire when the propeller blade was directly in line with the barrel.

The trouble was that, at first, they didn't work. A French pilot, Roland Garros, worked with aircraft designer and engineer Raymond Saulnier to perfect a device that could be fitted to his Morane-Saulnier aircraft in 1915. They fitted a Hotchkiss machine gun in front of the cockpit to fire forwards through the spinning propeller but, while Saulnier's device appeared to work, the propeller was still being shot to bits. Deficiencies in the machine gun or its ammunition meant that the bullets did not always leave the gun at the same rate or speed, and precise timing was vital. Garros, desperate for a usable weapon, decided to fit armoured propeller blades with deflector plates to ensure that any stray rounds flew off to the side rather than straight back at himself.

It worked. Garros scored several 'kills' before being shot down by gunfire from the ground. Dutch aero engineer Anthony Fokker installed a different type of synchronisation system on a Fokker Eindecker monoplane, also in 1915, and enjoyed far greater success. There was no need for an armoured propeller, which hampered the performance of the aircraft, and the German pilots began to dominate the air. It was almost a year before the British perfected their own system but all of these mechanical linkages, relying on the rotation of the propeller shaft to either lock or release the machine gun trigger, affected both the aircraft's performance and the machine gun's rate of fire.

A Romanian engineer named George Constantinesco eventually devised the system with which all British aircraft would eventually be fitted. Constantinesco was an expert in the theory of sonics and his synchronisation system involved energy being transmitted through a liquid, allowing for a much faster reaction than the purely mechanical systems, meaning that the aircraft's guns achieved almost their normal rate of fire and the engine performance was not affected. Even when British planes fitted with the system were shot down, the Germans could not work out quite how Constantinesco's system operated and it remained a secret until long after the First World War.

Machine guns firing through the propeller continued to be used on some aircraft all through the Second World War and even on some types in Korea in 1951, although the wing designs of fighter planes were, by then, well able to accommodate machine guns or cannon, making synchronised firing unnecessary. The advent of the propellerless jet fighter ultimately made synchronisation obsolete.

THE IRON MAN SUIT

(POWERED EXOSKELETON)

Anyone who has seen Robert Downey Jnr. defending the world as Iron Man in the movies based on the Marvel Comics character knows how hard Downey's character, playboy Tony Stark, had to fight not only against the bad guys, but against the military men who wanted control of the Iron Man suit.

The suit, a high-tech exoskeleton that serves as armour, houses weapons systems, gives the wearer superhuman strength and allows him to fly faster than a fighter jet, clearly has military potential and in the real world, the military has been looking for something just like it for years. Naturally, they have been prepared to start with something a little less ambitious.

The first device that could be described as an exoskeleton was built by a Russian engineer in 1890 and used compressed air to help put more power into the operator's natural movement. In1917 an American inventor came up with a more elaborate concept that used steam power. The technology really wasn't advanced enough for either design to work, either as equipment for a super-soldier or even to make it possible for manual workers such as warehousemen to lift heavier loads.

It wasn't until the 1960s that some serious research was undertaken to create a practical exoskeleton, General Electrics working with the US military to produce Hardiman, a contraption that looked like it was part man, part robot and part fork-lift truck. Using electricity and hydraulic systems, the suit could, in theory, allow a man to lift 25 times the weight he would normally be able to manage. Unfortunately, Hardiman's movements proved difficult to control and so violent that it was never tested with anyone inside.

Research into the feasibility of exoskeletons continued and in 1986 a former US Army Ranger named Monty Reed, who broke his back during a night-time parachute jump, was recovering in a hospital bed when he read the science fiction novel *Starship Troopers*. The troopers of the future had Mobile Infantry Power Suits that not only carried their equipment but would walk them off the battlefield to an aid station if they were injured. With no formal training, Reed designed his own Lifesuit that would help disabled people to walk. He is now on the 14th version of the Lifesuit prototype, having used them to take part in charity races – he completed a 3-mile event in 90 minutes – and he holds the world record for speed walking in a robot suit. Having funded most of his work himself, Reed's Lifesuit is a hugely impressive achievement and, although it still looks more Scrap Yard Man than Iron Man, he has made significant progress.

In Japan, the HAL-5 suit has been developed that uses the wearer's nerve impulses to assist movement and increase human strength by a factor of five, and in the United States, Raytheon is developing a suit with military backing that can multiply the wearer's strength by up to 17 times.

Iron Man is still a long way off but, for people like Monty Reed, the benefits that are starting to appear – from research that is turning military science fiction into medical science fact – have huge potential.

WHEN TO LEAVE
THE CAR KEYS
AT HOME

(THE ARMOURED CAR)

Just as wealthy aristocrats in cavalry regiments once took their own horses to war, Belgian Army officer Lieutenant Charles Henkart decided to use his own mounts when he formed a motorised reconnaissance unit during the early part of the First World War. Henkart had his two cars, an Opel and a Pipe (which was a Belgian make) fitted with armour plating and machine gun mounts. He and his men caused havoc amongst German cavalry units until they came off worst during a firefight in September 1914 and Henkart was killed.

The bold Belgian had certainly started something, though. Belgian luxury car makers Minerva soon offered a version of their touring car with armour plate and a machine gun position where their wealthy customers (who included the King of Belgium, Scandinavian royalty and even Henry Ford) once sat in supreme comfort and the Belgian Army formed a Corps of Armoured Cars.

The Belgians were not the first to deploy the car in this way. The Italians were the first to take an armoured car into battle during their war with Turkey in 1911. Isotta-Fraschini were better known for producing sports racing cars than military vehicles, but they produced a vehicle, covered in armour plating, with a machine gun mounted to be fired from the cabin and another in a rotating turret, that saw combat in Libya.

It was no accident that what we would regard as exotic brands were used as armoured cars. Rolls-Royce produced a number of armoured car designs (Charles Rolls had once owned and sold Minervas as a dealer). The fact was that the motor car, like the aeroplane, was really still in its infancy during the First World War and only the most expensive, best-engineered vehicles had the reliability required or the power to cope with being weighed down with armour. The Isotta-Fraschini could reach a creditable 37 mph as an armoured car and Lawrence of Arabia, who used a Rolls-Royce Silver Ghost for charging around the desert, believed that 'a Rolls in the desert is above rubies' and said his greatest desire would be to have '… my own Rolls-Royce car with enough tyres and petrol to last me all my life.'

With wheels instead of tracks, the armoured car developed into a vehicle that was more versatile than a tank, faster on roads yet still able to travel cross-country where the going was good, especially if equipped with four- or six-wheel drive. It was less expensive than a tank but also more vulnerable with less armour or firepower. The manoeuvrability of the armoured car made it ideal for scouting ahead of an armoured column or for use in an urban environment. An armoured car could race in and out of a village where a tank could become bottled up in the street. Specially designed armoured cars, rather than vehicles simply converted from civilian use, saw service in every area of conflict during the Second World War and have continued to play a vital military role ever since.

Because they are not as intimidating as tanks, having wheels instead of tracks and often recognisable as 'ordinary' vehicles, armoured cars are used extensively for internal security. Police and army units will use vehicles such as armoured Land Rovers during periods of civil unrest without fear of the situation escalating in the way that it might do if a tracked vehicle – clearly a military vehicle and seen to be aggressive – was to be deployed in an urban area.

While modern civilian SUVs and pick-up trucks are frequently pressed into service nowadays by irregular army units or paramilitary outfits, young army officers in most countries are no longer required to take their own cars into battle as Charles Henkart once did. And their insurance companies can probably be grateful for that!

FLYING
IN COMFORT
ABOVE
THE CLOUDS

(PRESSURIZED AIRCRAFT CABINS)

Sitting in a modern airliner, perhaps on a holiday flight looking forward to a relaxing couple of weeks in the sun, passengers are treated to what early aviators would have regarded as an implausibly luxurious experience. They have hot meals and drinks, movies to watch, air conditioning, comfortable seating, even beds on some long-haul flights. Most importantly, they are accommodated in a pressurized cabin, without which they would not have such an enjoyable flight.

The first aeroplanes with pressurized cabins, or at least pressurized cockpits, were military machines in the 1920s and 1930s that were adapted to be flown at high altitude, setting records in the process, but also for the purpose of investigating the effects that flying at great altitude have on the aircraft and its occupants. What became clear was that, at higher altitudes, where the air is thin, aircraft suffered from less drag, so the engines didn't have to work so hard to drive it through the air. This meant higher speeds and better fuel economy. There is also less turbulence at high altitude as the plane flies above the weather systems.

The Boeing B-17 Flying Fortress entered service with the US Army Air Force in 1938, three years before America became embroiled in the Second World War. It had a range of around 2,000 miles and could operate at an altitude of 35,600 feet (10,850 metres). At that altitude, were you flying over the Himalayas, you could look down on the top of Mount Everest more than a mile below you and the outside temperature can drop as low as -60° Celsius. There is also no breathable oxygen. The ten-man crew of a B-17 had to wear oxygen masks, several layers of clothing, electrically heated flying suits or electrically heated underwear and the famous Irvin sheepskin flying jackets as well as layers of gloves, socks, insulated boots – so much clothing, in fact, that moving about was not easy.

The same was true, of course, for aircrew flying in other heavy bombers during the Second World War and the experiences of military aircrew made aircraft manufacturers realise that, while they had developed aircraft that were suitable for carrying heavy loads over long distances, those loads could not include paying passengers unless drastic changes were made. It was not practical to pressure the bomber fleets, but it would be essential for commercial airliners. The aircraft had to fly at altitude for fuel economy and a smooth ride but paying passengers would not want to wear oxygen masks, let alone dress for Arctic conditions.

The first commercial airliner to have a pressurized cabin was the Boeing 307 Stratoliner, which was based on the B-17 heavy bomber. The Stratoliner could carry 33 passengers and the first aircraft – only ten were built – were delivered to TWA and Pan-Am in early 1940. While other passenger aircraft continued to operate at low altitude, the Stratoliner, using pressurization technology developed for military use, blazed the trail for modern air travel.

A BAD DAY
FOR THE GENERAL

(BAKER RIFLE)

Rifleman Thomas Plunkett had no reason to believe that his name would be remembered more than 150 years after his death. Ordinary infantrymen don't usually become famous. Plunkett, however, performed a feat of marksmanship that was unprecedented, and he did so using the Baker rifle.

Most firearms used by infantrymen on the battlefield in the 19th century, including the British soldier's 'Brown Bess' flintlock, had a smooth surface inside the barrel, which was basically just a tube to send the bullet in the right direction. A rifle was an improvement on the smooth barrel because it used grooves on the inside to make the bullet spin as it was fired. The spinning bullet had what is called gyroscopic stability, making it fly straight. This improved the range and the accuracy of the weapon.

The British Army had been testing rifles for a number of years, generally finding them slower to load than normal muskets and not so easy for a soldier to keep clean as the rifling tended to pick up more carbon deposits than the smooth barrel. Muck in the barrel would slow the bullet as it was fired, making it almost useless. In 1800, the army invited the finest gunmakers to show off their latest rifle designs and the Baker rifle, made by master gunsmith Ezekiel Baker of Whitechapel, London was chosen as the weapon to equip the new Experimental Corps of Riflemen. After agreed modifications had been made, an order for 800 Baker rifles was issued.

The Experimental Corps became the 95th Rifles Regiment of Foot, but the new riflemen were no ordinary soldiers. They were elite troops, not simple 'redcoats'. While other British infantry regiments wore the famous scarlet tunic, the riflemen wore dark green. They were not to advance in formation towards the enemy as other regiments did. Riflemen were trained to work in pairs, think for themselves and use their skills as marksmen to fight far more like modern-day soldiers. They were skirmishers who would scout ahead of the regular troops and snipers who would remain behind to cover a retreat.

It was while doing this that Thomas Plunkett earned his place in history. A Brown Bess musket had an effective range of between 50 and 100 yards. The Baker rifle was accurate up to around 200 yards. At Cacabelos in Spain in 1809, as the British Army retreated towards the port of Corunna with the French in hot pursuit, Plunkett took up a firing position to snipe at the enemy and shot the French General Colbert from a range estimated at 600 yards. Before dropping back to join his fellow riflemen, he took the time to reload and, proving his first shot wasn't a lucky hit, Plunkett shot a major who had gone to the general's aid. Without their general to lead them, the French advance faltered, buying the British some much-needed time.

The Baker rifle, with its high degree of accuracy, allowed soldiers to choose individual targets instead of simply participating in massed volleys of musket fire and, as more rifle regiments were formed, it changed the way that troops were deployed in battle.

CARRY ON SMILING

(INVISIBLE BRACES)

Tom Cruise might have thought it was 'Mission Impossible' when it came to straightening out his wayward teeth. His dazzling smile, after all, is his enduring trademark. He has been flashing his pearly white teeth at cinema audiences since he first shot to fame in *Risky Business* in 1983 (his previous movie roles gave him nothing much to smile about) but after almost 30 years of exposure he found that his smile was starting to show signs of wear and tear.

When most people suffer from wonky teeth, the answer is often to use braces which, gently but firmly, persuade your teeth to line up for inspection exactly as you would want them to. If you are Tom Cruise, however, and famous for your smile, or you simply care enough about your appearance not to want to look like a team of miniature scaffolders have been at work in your mouth, then you will be desperate to find something a little more subtle.

Thankfully, NASA provided the solution. In 1987, invisible braces became available to anyone who could open their mouth and their wallet wide enough. The material used to create the new braces was TPA – translucent polycrystalline alumina. A company called Ceradyne, which produces ceramic armour for fighter-pilot seats, body armour for the FBI, soldiers' combat helmets and a host of other products for military and civilian use, developed TPA in association with NASA. Their aim was to use it to form the nose cones of missiles, where various tracking antennae are housed, as the ceramic material would allow infrared and other microwave energy to pass through it in order for the missile to use its sensors to lock on to targets. While allowing such signals to pass through it, the TPA nose cone would still be tough enough to protect the sensitive equipment against the rigours of high-speed flight and the drastic temperature changes involved in soaring through the upper atmosphere.

While Ceradyne were working on noses (of missiles) another company, Unitek, was working on mouths. Unitek is a major supplier of dental products and was looking for a way to help teenagers and sensitive adults tackle the genuinely distressing problem of straightening teeth without using unsightly metal reinforcement. Because it was translucent, allowing light to pass through it, TPA was an attractive substitute for metal when it came to creating braces. The translucent material would simply take on the colour of the surrounding teeth, making it far less noticeable. Obviously, being used in the nose cones of missiles, TPA was also incredibly tough stuff and would be able not only to pull teeth into line but to survive the stresses and strains of daily use in the patient's mouth. More importantly, its strength would also allow it to be used sparingly, keeping the braces as small and unobtrusive as possible.

Since their introduction, invisible braces have been developed under a number of different trade names to the relief of all of us who want to see Tom Cruise, and anyone else who needs to keep their teeth in line, carry on smiling.

A CUPFUL
OF MILITARY
TECHNOLOGY

(FREEZE DRYING)

If you are gasping for a decent cup of coffee and happen to be in one of the few places on Earth where there is no street corner coffee house, no Starbucks and no shop where you can buy 'real' coffee, you might have to try 'instant' coffee. There is then a very good chance that your instant brew will be made from freeze-dried coffee, first marketed by Maxwell House in 1963 but made possible by the needs of the military.

Freeze drying involves first freezing a product and then introducing a partial vacuum to the freeze-drying chamber that draws water out of the product through sublimation, which is when water turns from a solid into a gas without going through the liquid stage. The water is then removed from the chamber and the temperature very slowly increased from -30° Celsius to room temperature, at a rate of just a few degrees each day. The end result is a product that can be stored at room temperature almost indefinitely.

In South America, people living in the high Andes Mountains discovered a crude form of freeze drying as a way of preserving food hundreds of years ago and, although scientists had dabbled with freeze-drying techniques since the late 19th century, it wasn't until the Second World War that the process was developed to successfully freeze dry in bulk. The need arose because medical products such as blood plasma needed to be transported from America to Europe to treat wounded service personnel. The medical supplies needed to be kept refrigerated but there was not enough refrigerated transport available to cope with demand. Plasma was freeze dried to a powder that was lighter and less bulky to transport but easily rehydrated when needed.

After the war, different food companies began looking at ways of using freeze drying and discovered that, because the process leaves the cell structure of the product intact, it can be used to preserve some foods without having an adverse effect on the taste. Freeze-dried coffee, for example, became an instant success, even though it's not everyone's cup of tea.

'BOMBS GONE!'

(BOMB SIGHT)

When you are flying four miles above the ground lying flat on your belly staring down through the transparent nose cone of a Lancaster bomber at a speed of around 200 mph at night with searchlight teams down below trying to catch you in their beams, anti-aircraft gunners taking pot shots at you and fighter planes buzzing around determined to blow you out of the sky, it's a stressful time to start trying to work out complex mathematical calculations. Add to that the fact that all of your six crewmates are waiting for your 'Bombs Gone' signal that means the pilot can turn the aircraft round and take you home, and even the coolest of number-crunching accountants is going to struggle.

The calculations that had to be made involved factors such as the speed and altitude of the aircraft, the wind speed and the type of bombs you were dropping. When a bomb left the bomb bay of the aircraft, it would be travelling forwards at 200 mph, the same as the plane. With no power of its own, its forward speed would quickly begin to drop, but its speed towards the ground would increase. The bomb would travel in a curve towards the ground and making that curve culminate in the target you were trying to hit was what all of the calculations were about. Fortunately, you didn't have to make those calculations. You had the Mark XIV Computing Bomb Sight to do that for you.

Bomb sights had been in use ever since aircraft started dropping bombs on the enemy during the First World War. At that time, the bomb sight might simply be a few lines drawn on the aircraft fuselage or a series of nails in a wing spar that you lined up with the target to let you know when to let go of the bomb that you were holding in your hand as you leaned out of the cockpit. More technical bomb sights came along later in the war.

By the Second World War, bomb sights had become far more sophisticated. The Course Setting Bomb Sight used by the RAF could take into account wind speed, a major factor with aircraft then able to fly so much faster and higher than ever before. Bombs had fins at the rear to stabilise them in flight and keep them on course. Flying into the wind, the bomb would fall straight, although its forward speed would drop off more quickly. The later Computing Bomb Sight used an analogue computer to make calculations that automatically adjusted the sight as the bomb aimer waited for the cross-hairs in the sight to creep over the target before he pressed the button to release the bombs.

The Americans had an even more elaborate bomb sight that allowed the bomb aimer to take control of the aircraft via the automatic pilot system in order to make minor course corrections to get his cross-hairs over the target.

On bombing ranges during training missions in good conditions, with clear skies and an easily identifiable target, bomb sights brought excellent results. Under operational conditions, things seldom went so well.

If the target was obscured by cloud or smoke from fires burning on the ground, the bomb aimer would not be able to line up the bomb sights. Under those circumstances, they would often bomb according to 'dead reckoning' which involved working out their exact position and dropping the bombs at what they calculated to be the right moment. A bomb being dropped from four miles high had to be dropped when the aircraft was more than a mile short of the target. Any small error would mean that it fell nowhere near the mark. If the wind changed direction at a lower altitude, the flight of the bomb could also be affected.

Coloured flares were often dropped by 'Pathfinder' aircraft at night to mark targets on the ground and give the following bomber fleet an aiming point. Defenders on the ground might extinguish those flares and light other flares elsewhere to lead the bombers away from the intended target.

There was also always the strong possibility that the aircraft's equipment might be slightly out, especially if it had come under heavy fire, which would really shake things up. A 5 per cent error in the air speed indicator, for example, could mean that the aircraft was travelling at 210 mph, or 190 mph, instead of 200 mph. A similar 5 per cent error in the calibration of the bomb sight could mean that a bomb falling from 20,000 feet might fall a third of a mile short of the target, or a third of a mile beyond it.

Reconnaissance photographs taken in the wake of RAF bombing raids on Germany in 1941 showed that only 10 per cent of bombs dropped were falling within five miles of the intended target and only 20 per cent of bombs dropped by the US Eighth Air Force on 'precision bombing' raids managed to fall within 1,000 feet (303 metres) of the target.

Although radio direction finding systems and radar ultimately brought about huge improvements in bombing accuracy, the poor performance of precision bombing was a major factor in changing the strategy of the air war to encompass area bombing, where a whole district or even an entire city was the designated target instead of an individual industrial or military complex.

This, of course led to the firestorms that destroyed vast areas of cities such as Hamburg, Dresden and Tokyo.

The bomb sight was a military invention that changed the world largely because, ingenious though they were, they never really worked.

THE COOL
MACREADY LOOK

(RAY-BAN AVIATORS)

John Macready was a Californian, an economics graduate from Stanford University, a pilot and one of the coolest dudes on the planet. Macready joined the US Army in 1917 at the age of 19 and trained as a pilot. He was one of the best, quickly becoming an instructor and a test pilot, setting several records for high-altitude flights and endurance flights. In 1923, along with co-pilot Lieutenant Oakley Kelly, he became the first man to fly non-stop across America from the east to the west coast.

Staying airborne for so long on a regular basis, Macready began to become concerned about the bright sunlight damaging his eyes. Pilots wore goggles when flying in aircraft with open cockpits to protect their eyes from wind rush and debris, and some goggles used smoked glass to combat the sun's glare. They were not, however, very comfortable and could restrict peripheral vision, the last thing that a fighter pilot wants when he needs constantly to be on the look-out for enemy planes. Neither did the smoked-glass goggles shield the pilot's eyes from damaging infra-red and ultra-violet light.

Following a long balloon flight that left him with serious eye strain, Macready approached the New York-based company

Bausch & Lomb to find a solution. The company had been making eye glasses since the middle of the 19th century and supplied items like field glasses, periscopes and searchlight mirrors to the US military. By the 1930s, modern aircraft tended to have closed rather than open cockpits, so the problems of wind rush and flying debris were no longer an issue. Bausch & Lomb worked on creating a lens that would cover as much of a pilot's line-of-sight as possible, adopting a teardrop shape. The glass lenses, designated G-15 by the company, were tinted green and filtered out both infra-red and ultra-violet light. Thin, lightweight metal frames ensured minimal restriction of the wearer's peripheral vision and gold plating the metal added a touch of style.

Bausch & Lomb called the glasses G-15 Anti-Glare, establishing a patent in 1937 under the name of their new sunglasses division, Ray-Ban. The sunglasses were an immediate hit with pilots and quickly became known as 'Aviators'. The Ray-Bans became an essential, highly treasured piece of kit for American pilots during the Second World War but were not widely available to the general public until servicemen coming home from the war brought their sunglasses with them. Suddenly everyone, from Hollywood stars to fashion-conscious teenagers, wanted Ray-Ban aviators, turning John Macready, although he probably never realised it, into one of the coolest fashion icons of all time!

FULL METAL JACKET

(CARTRIDGE)

About the size of a finger and light enough for one soldier to carry a supply that could eliminate an entire battalion, the metal cartridge transformed military firearms and changed the nature of warfare forever.

Prior to the middle of the 19th century, preparing a rifle or pistol to fire meant loading at least three separate elements – the gunpowder charge, a percussion cap to ignite the charge, and the bullet. The percussion cap created a spark to set off the gunpowder and took the place of the earlier system where a separate gunpowder 'primer' had to be lit to create a flash. Wadding in the form of paper or cotton was also packed into the weapon with the bullet to create a snug seal behind the bullet so that the expanding gas from the explosion of the gunpowder charge all went towards pushing the bullet out of the barrel rather than leaking out round the sides.

Inevitably, loading the individual elements into the weapon took time, limiting the soldier's rate of fire. Consolidating the elements into one piece of ammunition meant not only that a single round could be loaded more quickly, but that the process could be automated.

The modern metal cartridge, which began to be introduced in the late 1840s, quickly evolved into the ammunition used today which is different only in detail from its predecessors over the past 170 years. The ammunition used in a cannon or even an artillery gun is recognisably the same family as that which is used in the sidearm carried by police officers the world over.

The 'bullet' we have all seen, whether at the movies, on TV or in real life, is actually a cartridge that has a percussion cap at the base of a metal cylinder (the cartridge case), usually brass, that contains the propellant charge. Although still referred to as 'gunpowder', the original black powder is no longer used as it produced too much smoke and left residue that clogged the barrel and firing mechanism of the weapon.

At the end of the cartridge case is the cone-shaped bullet. This is held tightly in the mouth of the cartridge case, to be expelled with the propellant is detonated. The bullet may be solid metal (such as lead), a combination of metals inside a metal shell, or a combination of metal and explosive or incendiary material inside a shell. The shell is what gives rise to the term 'full metal jacket'.

Ammunition with this basic anatomy is robust and rigid enough to be inserted into a weapon mechanically and, while this first made a major impact in the revolvers and repeating rifles of America's Wild West, it also inspired the invention of the machine gun. And it was the machine gun, bringing the engineering advances and techniques of the industrial age to the battlefield, that used the metal cartridge to even deadlier effect.

INDUSTRIAL AGE WARFARE

(THE MACHINE GUN)

From submachine guns like the Thompson 'Tommy' Gun used by gangsters of the Al Capone era, to machine pistols like the Uzi used by gangsters of the *Grand Theft Auto* era and heavy calibre weapons like the Browning .50 Caliber, machine guns come in all shapes and sizes. The one thing that they all have, the thing that unarguably establishes them as a military invention that changed the world, is firepower.

Attempts to produce multi-shot weapons that would fire bullets in sequence, or sometimes all at once, had been made with reasonable success since the early part of the 18th century but they were limited both by the engineering technology and, more importantly, by the ammunition available. It wasn't until the metal cartridge came along that practical automatic firing systems could be developed. One of the first was the Agar Gun, also referred to as the 'coffee mill gun' because it relied on the operator cranking a handle as though using a coffee mill. Ammunition was fed into the gun from a hopper on top of the weapon, the crank acting as a trigger to fire the round and eject the spent cartridge case and re-load.

The Agar Gun actually used paper cartridges that had to be loaded into tubes that were then placed in the hopper and the tubes were ejected below the gun to be reloaded with paper cartridges and put back in the hopper. The gun crew certainly had their work cut out but their efforts were largely wasted as the gun had a limited range and was prone to the barrel overheating. It saw limited service during the American Civil War (1861-65) but was phased out by the American Army thereafter.

The next significant development in the machine gun was the Gatling Gun in 1861. Although later versions used metal cartridges, the first Gatling Guns used a system similar to the Agar Gun where a hopper or stick magazine fed steel tubes into the mechanism. The difference was that the Gatling Gun had six, or even as many as ten, barrels, each with its own firing mechanism, that revolved as the crank handle was turned. They fired when they rotated into position, ejected the cartridge case and were reloaded. Having so many barrels meant that overheating was not so much of a problem during sustained firing and the Gatling Gun could achieve a rate of fire of more than 400 rounds per minute.

Neither the Gatling Gun nor the Agar can really be considered as true 'automatic' weapons, because they had to be cranked, which is the same, in effect, as someone pulling a trigger to fire each individual round. The first fully automatic weapon was the Maxim Gun, invented by Sir Hiram Stevens Maxim in 1884. Maxim used the recoil force from the gun to operate its mechanism, which can be most easily imagined if you think of a farmer firing a shotgun. When the farmer pulls the trigger, the blast from the shotgun cartridge forces the expanding gas from the explosion to exit along the barrel, pushing a load of pellet out towards an unfortunate rabbit. But the gas and shot must push back against something to achieve their forward momentum – action and reaction, Newton's Third Law. The backward force travels through the gun to the farmer's shoulder, pushing his body backwards, which is why the sensible farmer will have been leaning forwards into his gun when he took aim.

Maxim used that backwards force, the recoil, to operate a spring-loaded mechanism that would travel backwards, eject the spent cartridge, re-cock the weapon and then load another round into position as its spring pushed it forwards again. As long as the machine gunner kept his finger on the trigger, the cycle would continue. Ammunition was fed into the gun on a belt that could hold 250 rounds, allowing the Maxim to fire 600 rounds per minute, roughly equal to 30 men with rifles.

The Maxim Gun usually had a crew of four who were needed to carry ammunition and the water that was used in a jacket around the barrel to stop it from overheating, as well as to feed the ammunition belt into the gun and to fire it. The Maxim Gun and its direct descendants changed the way that battles were fought and by the advent of the First World War in 1914 it quickly became clear that a troop of men advancing towards the enemy in an orderly line would be turned into a mass of bodies in the face of machine-gun fire. New tactics would be required to engage the enemy on the battlefield of the 20th century.

Later machine guns would become lighter and less cumbersome, using more refined mechanisms and ammunition, but even today they operate on the same principle as Maxim's gun. Some vent expanding gas from the barrel to operate the mechanism in slightly different ways but they still rely on Newton's Third Law.

All, that is, except for those that have returned to the Agar and Gatling technique with rotating barrels, now spun by electric motors. The new breed of 'Gatling' is generally of a larger calibre and mounted on ground attack aircraft or helicopter gunships.

LET THE SWEAT POUR OUT

(THE JUNGLE BOOT)

While the old adage about an army marching on its stomach certainly does hold water, a soldier most definitely marches on his feet and what he really doesn't want is boots that hold water. A regular leather army boot will keep out the worst of most kinds of weather but in the jungle they were a real problem.

In the hot, damp conditions in the tropics, when soldiers on patrol could spend most of their time trudging through swampy ground, regular boots proved to be a nightmare. When troops had to wade through water the boots filled up and held the water in. They became saturated walking across damp ground and with their feet sloshing around in sopping wet leather, blisters and sores quickly developed. In the humid conditions the leather also deteriorated, damp stitching rotted and a new pair of boots could very quickly become useless.

Prior to the Second World War, US troops operating in Panama tried out a new kind of boot. Instead of trying to keep water out, it was designed to let water out. Rubber soled and with a canvas upper, the boot had drain holes that let water and sweat drain while allowing air to circulate and help prevent fungal infections. The drains were protected with mesh to try to keep out insects and mud. Within a couple of years, the boots had been improved with inner soles that helped to pump moisture out of the vent holes and move fresh air around the inside of the boot.

During the Second World War, American jungle boots, or tropical combat boots, became highly prized articles amongst other Allied soldiers whose own footwear was often woefully inadequate. They remained highly prized throughout the 1950s, and further improvements were made during the Vietnam campaign including the insertion of a steel plate in the sole to protect soldiers from sharpened bamboo stake booby traps in hidden pits on jungle trails. By this time Australian troops serving in the region were even rumoured to be willing to swap their treasured 'slouch hats' for a pair of American jungle boots.

The successful development of the American jungle boot encouraged other countries to create their own versions and the idea that the poor old infantryman had to use one type of boot to suit all occasions became a thing of the past.

TORPEDOES IN THE SKY

(CRUISE MISSILES)

Journalists in Baghdad in 1991 could scarcely believe their eyes when they looked out of their hotel windows as the city came under bombardment during Operation Desert Storm. Tomahawk cruise missiles were flying past, following the layout of the streets, using their own guidance systems to seek out specific targets vital to Saddam Hussein's military command and control structure.

Foreign news reporters, who had remained in Baghdad to broadcast details of the assault that everyone knew was coming, ventured out from basement shelters to catch a glimpse of what was going on and were amazed. Veterans among them had expected to see or hear bombers overhead dropping their deadly payloads, but the unmanned cruise missiles streaking past lower than some rooftops was something they had not experienced before.

Yet cruise missiles, or at least the concept of the cruise missile, were far from being a new idea. The British had experimented with radio-controlled aircraft during the First World War with the intention of creating a missile that could be guided to its target, although the weapon was never developed to a stage where it could be deployed. In 1918, the Curtiss Sperry Aerial Torpedo was successfully tested in America. It was a small biplane powered by a Curtiss engine using a gyroscopically controlled autopilot system that had been developed by the Sperry Company of Brooklyn. After a certain period of flight, the engine cut out and the aircraft dived onto its target, delivering its high-explosive payload.

A similar system, but using a monoplane capable of flying faster than any contemporary fighter, was developed by the British during the 1920s as an anti-ship missile. It was tested at a range of over 100 miles but was difficult to control, inaccurate and unreliable. It wasn't until the 1930s, when both the Russians and Germans were working on rocket and pulse jet-propelled devices, that the cruise missile became a viable weapon. With Stalin having 'purged' some of his best men in the field after they were denounced as traitors by over-ambitious rivals, it was the Germans who were able to deploy the first cruise missile.

The German V-1 was not ready to be used operationally until 1944 and even then it suffered from reliability problems that saw 30 per cent of those launched against Britain fall harmlessly into the English Channel, although 2,419 fell on London. The other major V-1 target was Antwerp in Belgium which was hit by 2,448 of the missiles.

At the end of the Second World War, with the temperature of international relations plummeting towards the Cold War, both the United States and the Soviet Union developed cruise missiles that could be launched from land, from aircraft, from ships or from submarines. They used rocket boosters to set them on their way and then 'cruised' on jet engines. They could be armed with either conventional or nuclear warheads and were guided by radio remote control, the operator tracking the missile using radar. The development of these weapons was, however, overshadowed by the need for the superpowers to create Inter-Continental Ballistic Missiles (ICBMs) to deliver nuclear payloads.

Cruise missiles have, however, come into their own in the age of advanced electronics. The modern priority for a missile is not to take out an entire city as an ICBM would do, but to seek out and destroy a particular building or installation. Since the 1970s, cruise missiles have been developed that can find their own way to the target using a number of different systems including satellite navigation and radar that recognises the terrain over which they are flying. They can cover distances of more than 1500 miles and strike within 30 feet (less than 10 metres) of their target. Some fly low enough to avoid detection on radar, while others can fly at six or seven times the speed of sound, making them almost impossible to intercept.

Before the Second World War, in 1932, Britain's Prime Minister Stanley Baldwin warned that, in a future war, 'the bomber will always get through'. His words can be far better applied today to the cruise missile.

HERE COMES THE CAVALRY!

(CAROUSEL)

The most impressive part of any traditional funfair is, for anyone who enjoys a spectacle of lights, colour and sound, the carousel. While teenagers may dismiss the brightly painted galloping horses as being for little kids and turn their backs on the merry-go-round in favour of shoot-'em-up games in a video arcade, they really don't know what they are missing. The carousel has a history of real warfare that most video games can never hope to match.

The idea of the carousel was imported to Europe from the Middle East where Arabian horsemen used a similar set-up to practise their cavalry skills. Sitting in baskets or on saddles hanging beneath a tall pole, they would be spun round at speed to practise lopping the heads off dummy enemy warriors as they flashed past. The device may have been used up to 2,500 years ago and remained in use in Europe throughout the Middle Ages with wooden horses hanging from an arm that revolved, while balanced on a central pole riders used lances to spear rings, again as practice for going into battle.

The name 'carousel' was used for a kind of riding display set to music, cavalrymen putting on this show of skill to impress visiting dignitaries at state functions. They dressed in colourful uniforms and carried blazing torches to illuminate the scene as they trotted their horses in carefully rehearsed formations.

It's easy to see how the training apparatus and the military tradition eventually combined to become a popular amusement, so next time you stand watching the carousel go round, watch out for those cavalrymen galloping by!

THE TERRIFYING SOUND OF SILENCE

(PULSE JET ENGINE)

We see jet engines powering airliners through the skies above our cities every day of the week. We've become used to the sight and sound of them, although anyone living directly under a flight path to a major airport would love to find that, one day, the noise of the engines suddenly stopped. For anyone living below the flight path of the aircraft powered by the first jet engines, the sudden lack of engine noise was the last thing they wanted to hear.

The first working jet engine was not the sort of jet turbine that powered the first jet fighter. It was a far cruder device called a pulse jet. The pulse jet had first been proposed by a Russian artillery officer as long ago as 1864 but it was another Russian, an engineer named Karavodin, who produced the first experimental pulse jet engine in 1907.

A pulse jet engine takes in air at the front end, injects fuel into the air and burns the mixture, the resultant expanding hot gas exiting through a tube at the rear to generate thrust. A jet turbine, which is what most of us think of as a jet engine, operates on much the same principle as the pulse jet but does so on a continuous cycle as opposed to the pulse jet producing a series of burns in rapid succession. The jet turbine also compresses the air to produce greater expansion and more power from the burn, whereas the pulse jet uses minimal air compression prior to ignition. This makes it less powerful than a turbine but also a lot less complex.

While engineers in a number of countries were studying the nature of pulse jets, Russian research on pulse jets and rocket-propelled missiles stalled during the 1930s when Stalin 'purged' some of his finest scientists and it was Germany that took the lead. In 1933 an engineer named Paul Schmidt submitted designs for a pulse jet engine to the German Air Ministry and, with government support, Schmidt and designer Georg Madelung proposed a pulse jet-powered missile that showed promise but needed further development. Schmidt took the project forward, combining the efforts of several German companies, including aircraft manufacturer Fiesler and engine maker Argus. The result was the Argus As 109-014 engine, installed in the Fieseler Fi 103, better known as the V-1 flying bomb.

The simple pulse jet was cheap enough to manufacture for it to be considered expendable as part of the new missile. It could send the V-1 hurtling through the sky at 400 mph, as fast as the finest fighter aircraft of the day, and it delivered a warhead packed with a ton of high explosives. The pulse jet was not powerful enough to get the V-1 off the ground, take-off being made with the assistance of a ramp and a steam-powered catapult, but once airborne it performed admirably. The first powered flight took place towards the end of 1942.

The V-1 had a gyroscopic compass and autopilot system that kept it on course and a tiny propeller-like device in the nose called a vane anemometer. The propeller blades were turned by the wind rush as the V-1 flew through the air and once the blades had completed a pre-determined number of revolutions, the missile was calculated to be over its target. The V-1 then went into a steep dive which interrupted the fuel flow, causing the engine to cut out.

With the engine burning its pulse of fuel and air 50 times a second, it created a distinctive rasping sound that led to it being nicknamed the 'buzz bomb'. Approximately 10,000 V-1s were launched against England, most of them targeting London. Over 2,400 made it through the anti-aircraft and fighter defences causing 6,000 fatalities and 18,000 serious injuries.

Londoners soon came to know that, while the noise of the buzz-bomb's jet engine was annoying, hearing it stop was much worse, as it meant that a one-ton bomb was heading their way.

TOOLBOX
IN YOUR POCKET

(THE SWISS ARMY KNIFE)

If you've ever had to open a bottle of wine while digging a stone out of a horse's hoof and trimming your moustache, then the tool you needed must surely have been a Swiss Army knife.

The iconic red-handled pocket knife bearing the shield with the white cross of Switzerland is instantly recognizable the world over and, though you might think that it has very little military ancestry, it really was first devised for the Swiss Army. In the late 1880s, the Swiss Army decided that their soldiers needed a pocket knife that could also serve as a screwdriver to be used by the soldiers in servicing their rifles. It also had to be capable of opening canned food.

The knife that was accepted into service in 1890 included a blade, a screw driver, a can opener and a tool for making holes. Its grip was dark wood rather than bright red and 15,000 were ordered – to be produced by a German manufacturer. It quickly became known as the Soldier's Knife.

In 1891 a genuine Swiss manufacturer took over production of the knife, using a special spring arrangement in the handle that allowed for tools to be attached on either side. A small blade was added as well as a corkscrew. It was around this time that the knife also acquired its cross-and-shield emblem. The new version of the knife was not for general issue to the army and was known as The Officer's Sports Knife.

Production continued through the early part of the 20th century, with the knife proving increasingly popular with the general public. Then, after the end of the Second World War, US servicemen, having bought the useful, gadget-laden, pocket knives in large numbers at their PX stores, christened it the Swiss Army Knife, the German *Schweizer Offiziermesser* proving too much of a mouthful.

The name only added to the appeal of the knife and today there are numerous versions available in a range of colours and finishes and with enough gadgets to handle any tricky job you need to tackle when the only resource you have available is the toolbox in your pocket.

'UP PERISCOPE!'

(THE PERISCOPE)

The image of the submarine captain squinting into his periscope to search for a target is one that has lingered on cinema screens ever since the tension and claustrophobic terror of submarine warfare was first dramatized for the movies during the First World War. Dozens of submarine movies have appeared over the past century or so and war film fans would have been sorely disappointed to sit down and watch a submarine action flick that didn't include the command 'Up periscope!'

A periscope is essentially a long tube that uses angled mirrors to reflect the light coming in through the viewing hole at the top, down the tube and out via an eyepiece at the bottom. This allows anyone looking through the eyepiece at the bottom to see whatever they would see if they were looking out through the viewing hole at the top. Some devices use prisms instead of mirrors and some have lenses added to allow for a magnified image, but the principle remains the same.

The periscope may first have been used as a kind of novelty or toy and in the 15th century small periscopes are known to have been used by people in Aachen to see what was going on at a religious gathering, so vast and tightly packed were the crowds.

The military applications for the periscope were, however, quite obvious. Anything that would allow you to remain safely behind a wall or a parapet and see what the enemy was doing without him being able to shoot at you was clearly a good thing.

The simple novelty was developed for military use with lenses and range finders that allowed artillerymen to see where their shells were falling without getting their heads shot off or, when the tank first appeared during the First World War, tank commanders to take a look outside without opening any chinks in their armour.

Towards the end of the 19th century, submarine designers adopted periscopes that allowed the submarine captain to see what was happening on the surface while his boat was still submerged. At first these were fixed tubes but by 1902 American naval architect Simon Lake had devised a telescopic periscope that could be raised and lowered as required. Using hydraulic systems, this could ultimately be done at the flick of a switch, so we have Mr Lake to thank for the 'Up periscope' command.

The periscope was to become the submarine's greatest asset, turning it from a stealth ship, capable of small-scale subterfuge or sabotage, into an absolute predator preying on enemy shipping, stalking unseen beneath the waves and capable of striking before its target even knew it was there.

And yet, for anyone considering making a modern submarine movie, the 'Up periscope' dialogue line should probably not feature in the script. The latest submarines use photonics masts which are less vulnerable to springing a leak than the hollow tube of a periscope. Instead of mirrors, the photonics mast houses electronic sensors and sends digital images direct to screens inside the submarine.

GETTING
YOUR WIRES
CROSSED

(THE WIRE-GUIDED MISSILE)

The problem with shooting at a tank is that it doesn't always stay in exactly the same place and, if you are firing a missile from a safe distance of well over a mile away, by the time your missile reaches the spot you aimed at, the tank has moved on. Worse still, firing the missile could give away your own position and you end up dealing with an angry tank.

The same is true of aircraft and ships. During the Second World War, the Germans developed different systems for guiding missiles onto their targets. They used radio control in anti-shipping glider bombs but with only limited success. Launched from an aircraft, the bomb had to be 'steered' towards its target by an operator in the launch aircraft, which had to fly straight and level, allowing the operator to keep the bomb in sight and on course. The launch aircraft was easy prey for any fighters in the area and the Allies, who knew a thing or two about radio remote control, used jamming transmitters to create interference and make the operator lose contact with the bomb.

To avoid being jammed, the Germans tried using wire-guided missiles. These could send signals that could not be blocked, so the operator could keep the missile heading on target. This proved almost impossible for fighter pilots aiming to shoot down bombers as flying a fighter while controlling a missile required two sets of eyes and at least three hands. The missile had to be fired from an aircraft with more than one occupant and, although it was tested, it never reached the stage where it was used in combat. A version of this missile that could be used against tanks by an operator with his feet firmly on the ground attracted a lot of attention at the end of the war. The Germans never managed to put it to use on the battlefield, but others took the idea and turned it into a formidable weapon.

Generally controlled using a joystick device, the missiles had small flares on their fins that helped the operator to keep them in sight and on target. The control wire unwound as the missile flew through the air, trailing behind to a maximum range of up to two miles. Wire-guided anti-tank missiles were widely deployed from the 1950s onwards and, in 1973 during the Yom Kippur War when Egypt and Syria fought a 19-day battle with Israel, the desert was said to be criss-crossed with the wires of the Russian-built 'Sagger' missiles that were used to destroy an estimated 1,000 Israeli tanks.

Even though more modern laser-guided or self-guided 'fire and forget' missiles are now in use, wire-guided missiles are still in service today, effective over far greater ranges than early models and using guidance systems where the operator simply keeps the target in his sights rather than having to keep an eye on what both target and missile are doing. That way he's less likely to get his wires crossed.

THE RIPCORD RELEASE

(THE FREEFALL PARACHUTE)

Floating to earth beneath the billowing sail of a parachute is something that some people do for sport, some soldiers do as part of their job and an unlucky few have had to do as a matter of emergency. All have a host of aviation pioneers to thank for their experience.

Parachutes, or parachute designs, have been around in one form or another since the 15th century but it wasn't until the 18th century that devices such as French physicist Louis-Sèbastian Lenormand's wooden-framed, umbrella-like parachute were shown to work effectively. Bulky parachutes like Lenormand's, which might seem more like a kind of hang glider to the modern eye, were superseded by more lightweight silk designs by the end of the 18th century but these generally involved holding the entire parachute array in your arms before you jumped from the balloon, building or wherever else was your starting point, then letting it go as you fell through the air. It wasn't until 1906 that American Charles Broadwick, who performed parachute jumps for the amusement of crowds at fairs, came up with a way of folding a parachute into a backpack that the parachutist strapped on using a harness system.

Broadwick also devised the static line, a cord attached to his balloon that would pull the parachute out of the pack as he fell and snap when the parachute deployed, leaving him free to float downwards.

Broadwick's static line idea became the standard way for artillery observers, sent up in balloons during the First World War, to exit the balloon basket when their balloons were shot to pieces and during the Second World War static lines were used for paratroops exiting aircraft. It is still the method used today by recreational parachutists who are not indulging in a freefall jump.

Following the First World War, with military aviation fully established as a major part of the armed forces, the US Army began looking at ways to bring together the best of the different parachute technologies to develop a system that would give aircrew the best chance of escaping from a doomed aircraft. American inventor and graduate of Oxford University, Solomon Van Meter had invented a parachute backpack with a ripcord release in 1911. Van Meter's system, along with ideas from balloonist and US Army instructor Albert Stevens, were incorporated into a parachute that was tested for the army by Leslie Irvin in 1919. Using a ripcord meant that anyone escaping from an aircraft that was damaged could fall well clear of the plane before deploying a parachute. Jumping from a plane flying at 1,500 feet (454 metres), Irvin fell for 500 feet before pulling the ripcord. This released a pilot chute that pulled the main parachute from the backpack, allowing Irvin to drift the remaining distance to the ground. He unfortunately broke his ankle on landing, but was otherwise safe. This was the first freefall parachute jump using a ripcord system.

Irvin later formed the Irving Air Chute Company, with his name apparently mistakenly misspelt in his own company title, and by the beginning of the Second World War, Irving parachutes were in use in over 40 different countries. It is estimated that Irving parachutes saved 10,000 aircrew during the course of the war.

WHAT'S COMING NEXT?

(STEALTH AIRCRAFT)

In late 1944, flight tests were conducted in Germany on the jet-powered Horten Ho 229 flying wing aircraft designed to elude British radar systems, bypassing the entire air defence system of the United Kingdom. The aircraft was designed to fly higher and faster than contemporary fighter planes and in tests it appeared to perform very well. With the end of the war in Europe only a few months away, however, the jet bomber that was invisible to radar had simply come too late. Nothing quite like it would be seen again until 1988.

The Ho 229 was built with a steel centre structure but with wings made using plywood panels impregnated with carbon which the designer hoped would absorb radar waves rather than bouncing them back to the radar station where the aircraft's presence would be registered. The flying wing configuration, with no tail plane sticking up and the two jet engines incorporated into the fuselage, gave the aircraft a very low profile, also intended to make it difficult to spot on radar.

It was 44 years later that the US Air Force admitted to the existence of the Lockheed F-117 Nighthawk, the first operational stealth aircraft. During the 1970s, using computer-aided design techniques, aero engineers and designers at Lockheed's top secret Skunk Works in California worked out that an aircraft using flat panels could be given such a low radar profile that it would be virtually invisible to most systems. Instead of a sleek, streamlined aeroplane, they came up with an ugly wart of an aircraft with unsightly angles which, like magic, radiated radar waves away from the sender rather than bouncing them back. The skin of the aircraft was painted black using carbon to absorb radar, the engines were mounted on top of the wings to minimise the chance of their glow being seen from the ground, and fuel tanks and weapons were carried internally to avoid them creating their own radar blip.

In 1989, when US forces invaded Panama, the F-117 was used operationally for the first time but few people even knew about it. Although the aircraft had been in service since 1983, and its existence had been acknowledged in 1988, not many people had actually seen one until they were allowed to be filmed during the Gulf War in 1991. By then the larger B-2 Spirit stealth bomber was also in service, looking rather like the old Horten Ho 229 of 1944.

Tests on the Horten in 2008 using a full-scale model showed that it could have flown against Britain undetected until it was too late for the anti-aircraft defences to do anything about it. In other words, the stealth technology developed in 1944 would have worked. We know that the stealth technology developed in the 1970s worked. The question now is, what is currently being developed that we still don't know about?

RUBBERLESS RUBBER

(SYNTHETIC RUBBER)

Of all the natural, renewable products that we see and use every day, rubber is one of the most fascinating. The raw latex that goes to make rubber products is still collected in the same way that it has been for more than 2,000 years, with small amounts of sap bled from the bark of a tree in a tropical forest, collected in the morning when the sap pressure is high. It's a slow and laborious process, even when conducted on a massive scale on plantations of thousands of trees.

The greatest consumer of rubber is the car industry which takes 60 per cent of natural rubber production, mainly for use in tyres. Rubber is also used in other industries producing household goods from flooring to window frames but it is in equipping our cars with tyres that we use most rubber and that has been the case ever since the car began to be mass-produced at the beginning of the 20th century. Other machines besides cars, of course, needed to use rubber products and the coming of the industrial age led to a massive increase in demand. Less than 100 tons of rubber were produced in 1824 but by 1903 that figure had risen to more than 40,000 tons.

Supplies of natural rubber simply could not keep up with demand and in 1906 a German chemical company called

Elberfelder Farbenfabriken challenged its employees to come up with a viable synthetic rubber, offering a substantial cash prize if it could be done within three years. Chemist Fritz Hofmann beat the deadline by just a few weeks, perfecting a polymer called methyl isoprene, isoprene being a by-product from refining oil. At around the same time, Russian scientist Sergei Lebedev came up with another synthetic rubber made initially from alcohol produced from potatoes.

The new synthetics paved the way for the production of synthetic rubber on an industrial scale, as the First World War exacerbated the ever-increasing demand for natural rubber by creating shortages caused by disruptions to supply. Yet the synthetic rubber was not as durable as its natural counterpart, making it less desirable for tyres, and after the war the man-made rubber struggled to compete. Improvements to synthetic rubber and its manufacture were made during the course of the following 20 years, the Russians moving on from using potatoes to using petroleum, but the onset of the Second World War galvanised synthetic rubber production when the Japanese conquered Asia, taking over most of the world's natural rubber plantations.

America needed rubber for military vehicles, aircraft tyres and a host of other applications without which it would be impossible to go to war. The US government brought together the best chemists in the field and they came up with Government Rubber Styrene (GRS), the backbone of American synthetic rubber production during the war. Soon there were 50 factories across America manufacturing synthetic rubber in quantities far greater than the annual production of natural rubber in the whole world before the war.

Synthetic rubber and natural rubber are both used today for much the same purposes, although there are now far more man-made varieties specifically tailored to suit individual industrial requirements and, just as in the Second World War, production of synthetic rubber far outweighs that of the natural type.

HOW REMOTE IS REMOTE CONTROL?

(UAV)

If you have listened to a news broadcast or read a paper recently, you can't have avoided hearing about 'drone strikes' and UAVs (Unmanned Aerial Vehicles). These remote control or self-piloting aircraft are now used not only in war zones but also by security services and police forces for surveillance and even monitoring traffic conditions.

These aircraft come in all shapes and sizes, from devices that look like model aircraft and can be launched from your hand or spider-like helicopter machines using four blades supporting a camera platform, to high-altitude snoopers that can stay on station for days and the sinister hunter-killer UCAVs (the C standing for combat).

Drones that are used in missile strikes are equipped with highly sophisticated sensors including thermal cameras and long-range, high-definition video cameras that can read a car number plate at a range of two miles and relay images to the operator via satellite links. Hunter-killer drones like the United States Air Force (USAF) MQ-9 Reaper are as big as a small aeroplane and have been used extensively for air strikes in Afghanistan and several have been lost in action. The equipment they carry is still ranked as top secret and a drone will be destroyed rather than being allowed to fall into enemy hands. When a Reaper operated by the RAF crashed in Afghanistan, an SAS team was sent in to locate the wreckage, recover sensitive instruments and destroy the rest. In 2009, control of a USAF Reaper over Afghanistan was lost and it started flying towards the border with Tajikistan. A jet fighter – with a pilot on board – was despatched to intercept and destroy it.

Yet remote-controlled drones are nothing new. Experiments were being conducted with remotely piloted aircraft a century ago, and during the Second World War the RAF had a fleet of around 300 radio-controlled De Havilland Tiger Moth biplanes. These were used as flying targets for gunnery practice and designated DH.82 Queen Bee. It's from the Queen Bee name that the term drone, which is actually a male bee, is thought to come when applied to remotely controlled aircraft.

The RAF personnel controlling the Queen Bees did so using a radio set that operated pneumatically driven controls sited in the rear cockpit. The front cockpit had normal controls so that the Queen Bee could be flown normally when required. They had to keep the Queen Bee within reasonable range of their transmitter and the Queen Bee's receiver, and keep the aircraft in sight while obviously staying well clear of the firing line themselves.

The operators remotely controlling Reapers in Afghanistan really couldn't be more remote. They are stationed in a hangar at Creech Air Force Base near Indian Springs in Nevada, flying the Reapers via satellite.

THE SUBTLETIES
OF EAVESDROPPING

(LISTENING DEVICES)

Eavesdropping on other people's conversations when they don't know that you're listening is regarded as rather rude, but everyone knows it can be hugely addictive. Overhearing a couple chatting on a bus can be every bit as good as listening to a radio play and there's always that frisson of excitement in knowing that you're doing something a bit naughty.

Listening in on people covertly is by no means a new thing. People have been spying on each other this way for centuries, using holes in walls, listening tubes or simply putting an ear to the door of a room. The invention of the telephone and microphone in the 1870s meant that microphones could be used to 'bug' a room, although even once they could be made small enough, hiding a microphone to obscure the wires but ensure that it could still pick up speech was never easy.

The first truly successful eavesdropping device was nicknamed 'The Thing' and had no wires and no battery, making it very difficult to locate. It could be activated by the operator sending a radio signal at a certain frequency and would then basically reflect that signal back to the sender having modulated or altered it using the vibrations from sound picked up in the room where it was located. That room was the study inside the US Ambassador's Residence in Moscow. 'The Thing' was inside a carving of the Great Seal of the United States that was presented to the ambassador by a delegation of children from the Young Pioneer Organisation of the Soviet Union in 1945 to celebrate the friendship between the two Second World War allies. The bug remained hidden and operational until it was discovered by accident in 1952 when a British radio operator picked up American voices when scanning the airwaves.

The Americans and the British were very soon spying on the Soviets with their own versions of 'The Thing'. Making bugs that could transmit conversations became something of an art form as advances in electronics allowed for ever greater miniaturisation, but electronic equipment designed to locate electronic bugs has become just as sophisticated. It was reported that two bugs were found in a room in which British Prime Minister Tony Blair was scheduled to stay during a visit to India in 2001. Blair's security staff had him moved to another room.

Listening devices are used today in all sorts of ways. The smart phone you carry with you can be programmed with an app that will give away everything you say, without you knowing a thing about it. Invisible laser beams can be bounced off the window of a room, picking up vibrations from conversations inside and reflecting them back to a receiver.

Security services and police forces all over the world now have more ways at their disposal to eavesdrop on other people's conversations and, unlike someone listening in to a couple on the bus, they don't think it's naughty in the slightest!

SOME DAY MY PRINTS WILL COME ...

(THE DIGITAL CAMERA)

There used to be something special to look forward to when you came home from a holiday, something that provided an instant reminder of the wonderful time you had had, just as the entire holiday was starting to be crushed into a dark corner of your memory under the weight of everyday humdrum. The special thing was seeing your holiday snaps.

At one time, unless you had the right kind of equipment and were skilled enough to do it yourself, you had to remove the film from your camera and either take it to a camera shop or a chemist to have it processed, or mail it to a photo lab. The whole process generally took at least a week, and until your prints were mailed back to you, you had no way of knowing whether you had ruined your holiday photos by having the camera on the wrong setting or if you accidentally exposed the film to light when you opened the camera to take it out. And your roll of film would only have 24 or 36 pictures on it ...

Nowadays, when you can take and store more photographs on your mobile phone than most people used to take in a lifetime of holidays, it seems a bizarre notion to have to wait to see your pictures. But if you think that would be frustrating, imagine how military strategists felt when they needed to see pictures of enemy installations. During the Cold War, satellites circling the earth took photographs using film, but the military problem was not finding the time to take the film to a camera shop – it was getting the film back from outer space. Satellites had to eject canisters containing the film which, providing that it survived the heat of re-entry into the earth's atmosphere, then had to be snatched out of mid-air by a passing plane as it floated down on a parachute. Only then could the film be processed.

This problem was eventually overcome by building satellites that could develop their own film, scan the images and transmit them back to earth, but the useful spying life of the satellite was still defined by how much film it could carry. That was all to change in 1976 with the launch of KH-11, an American satellite designed to gather military intelligence, as so many of its predecessors had been. The difference with KH-11 was that it used digital imaging. Its photographs had a resolution equivalent to 0.64 megapixels – far less than you have on your mobile phone – but it could go on taking and transmitting pictures until the cows came home.

The first digital cameras for the commercial market, that could be made small enough that they did not need to be housed in something the size of a satellite, began to appear in the late 1980s and the market for film processing disappeared almost overnight. Nobody wanted to wait for their holiday snaps any more.

OPERATION CHASTISE

(THE BOUNCING BOMB)

The story of the Dambuster Raid in May 1943, when Lancaster bombers of 617 Squadron – known forever after as the 'Dambusters' in honour of the men who staged the raid – has been told and retold over the years. The brave young men who trained for and flew the missions that destroyed the Möhne, Sorpe and Edersee Dams have gone down in history with their exploits even known to the younger generation of the 21st century. *The Dambusters* is now a smart-phone video game app and the famous scene in Star Wars where the Rebel Alliance pilots attack the Death Star is based in part on Operation Chastise – the Dambuster Raid.

The object of the raids was to destroy the dams and their capacity for producing hydro electricity to power the German factories in the valleys below them. In cracking the dams, the water released would also cause extensive flooding that would also disrupt industrial output. The Germans knew that their dams would be important targets for RAF bombers and their anti-aircraft defences were formidable. Dropping bombs directly onto a dam from above required the kind of precision bombing that was proving impossible at the time and dropping torpedoes that could swim through the reservoir to detonate on the dam was made impossible by torpedo nets stretched across the water.

Scientist and aircraft designer Barnes Wallis came up with an idea that worked on the same principle as a stone skimming across a pond. He designed a 'bouncing bomb' that could skip across the surface of the reservoir, avoiding the torpedo nets until it hit the top of the dam whereupon it would sink to a certain depth and detonate. The bomb, originally envisaged as a sphere, was ultimately developed as a canister like a large oil drum. The aircraft carrying it needed a special bomb-release mechanism that would start the bomb spinning backwards before it was released at a precise height, while the aircraft was travelling at a certain speed an exact distance from the dam.

Naturally, this required brave and skilful flying with the crews fighting their way to the target through a hail of anti-aircraft fire in order to deliver their bomb at exactly the right moment. It wasn't the sort of thing that bombers during the Second World War were very good at, but they managed it. This was in spite of the fact that, of the 19 Lancaster bombers that set out on the raid, only 11 returned.

The damage that they caused was, at least in the case of the Möhne and Edersee dams, significant, but not totally catastrophic. There was great loss of life when huge walls of water roared down the valleys, wrecking homes and property but the dams were repaired within five months. The greater effect that the raid had on the course of the war, and the way that Barnes Wallis's bouncing bomb changed the world, was in showing Britain's allies that, despite having been an island under siege for three years, she was still capable of striking back at the enemy. That was important given that the Allied invasion of Europe would happen in a year's time. The other important factor was to prove that precision bombing was possible, leading to the development of Barnes Wallis's deep penetration 'earthquake' bombs that were used against U-boat pens, V-2 rocket sites and the German battleship *Tirpitz*.

ALL'S FAIR
IN LOVE AND WAR

(NYLON)

American troops started arriving in Britain in small numbers in January 1942 just weeks after the United States was drawn into the Second World War. Over the course of the next three years, nearly three million American personnel would be stationed in or pass through the United Kingdom.

What they brought with them was manpower, equipment, armaments ... and nylon stockings. Nylon had been invented by the DuPont company in America in the early 1930s and they were in a position to begin marketing it by 1938, when one of the first applications was in toothbrush bristles. The intention, however, when developing nylon, had always been to try to find a synthetic substitute for silk. With war clouds gathering and relations between America and Japan deteriorating, finding a silk substitute became a matter of urgency. The Americans knew that silk supplies from the Far East would cease if the Japanese seized control.

Why was silk so important? Prior to the war, the vast majority of fabrics were manufactured from cotton, with wool making up most of the rest and silk only a small percentage. Yet silk had many virtues, being both light and strong. Vast quantities of silk were used for making parachutes and from the moment that the

Second World War began in September 1939, parachutes were in great demand.

Silk was also used for aircrews' thermal underwear and in making flak vests, but nowhere was the need for silk greater than in parachute production. Silk was the only material that was light enough to use as a parachute while remaining strong enough not to rip under the strain. One parachute required no less than 65 yards of silk cloth. A B-17 Flying Fortress had a crew of ten men and each of them required a parachute. On the Schweinfurt Raid in October 1943, 291 B-17s went out and 60 of them did not return. The need for parachutes and replacement parachutes was enormous.

Nylon stockings, as an alternative to silk stockings, were unveiled at the New York World Fair in July 1939, with groups of women invited to try them. They were a huge hit and later that year nylons went into production, DuPont selling 64 million pairs in 1940. After that, however, nylons became scarce as industrial production of nylon was given over to military requirements. The volume of stockings that had been produced proved that nylon could be manufactured on an industrial scale and the new material went to war not only in the form of parachutes but also in ponchos, tents, ropes and many other items, including tyres.

Nylon stockings never made it to the UK officially until well after the war and silk stockings had become so scarce that fashion-conscious women had taken to having their legs dyed and stocking seams drawn up the backs. If young ladies in Britain wanted to dress like the movie stars they saw at the cinema every week, they simply had to have stockings and they knew that the glamorous American troops could sometimes supply them. If he could get hold of them from the United States, for a GI stationed in Britain nylons were one of the most potent weapons in his armoury …

THE JET
THAT JUMPS

(VTOL JET)

There was something of a rivalry between pilots who flew fast jets and pilots who flew helicopters. The two disciplines, of course, require different skill sets and the aircraft are designed for different purposes. When a jet pilot taunted a helicopter pilot by saying 'I have to slow down an awful lot just to fly twice as fast as you,' the helicopter pilot would generally respond with, 'Maybe, but you can't land in my back garden.' That conversation was turned on its head by the arrival of the jump jet.

From the very beginning, when the Wright brothers were bravely clinging on to their ingenious contraption on the beach at Kill Devil Hills in 1903, one of the major problems in getting an aeroplane off the ground had always been keeping the whole thing stable for long enough to get up enough speed across the ground for the airflow over the wings to provide lift. How much easier would it be if you could simply leap into the air?

Vertical take-off became a reality with the autogyro, a kind of cross between an aeroplane and a helicopter, in the 1920s and the helicopter in the 1930s but, while using so much of the available engine power to stay aloft, they weren't so good at covering long distances quickly. Some highly imaginative and ingenious feats of engineering were attempted, some using engines housed in wings that could rotate from the horizontal to the vertical. One example of such a tilt-wing aircraft is the V-22 Osprey (where the rotors rather than the wings tilt) which is used successfully by the US Marine Corps today. The first viable combat jet that had the capability to take off and land vertically, however, was the Hawker Harrier.

The Harrier's layout was similar to many jet fighter planes. Its engine was positioned longitudinally with the air intakes sucking air in at the front and the jet exhaust providing forward thrust. As a conventional jet it could achieve speeds approaching supersonic flight and carry a weapons payload that made it an ideal ground attack aircraft. But the Harrier was not a conventional jet. Its Rolls-Royce Pegasus engine delivered the thrust through nozzles that could be positioned horizontally, vertically or at any angle in between. Positioned vertically, the engine could lift the Harrier straight into the air. This made it ideal for use from forest clearings, the Harrier's intended role being as a front-line fighter facing up to the Soviet Bloc on the European Cold War battlefield. A jet that could pop up out of a hiding place in a wood rather than having to use a mile-long tarmac runway, effectively giving it the capability to operate from just about anywhere, was a major military asset.

The Harrier entered service with the RAF in 1969 and was also modified as a carrier-borne fighter for the Royal Navy. Although they have been retired from the British armed forces, Harriers are still operated by the US Marines as the AV-8. The Soviet Union developed its own Harrier lookalike, the Yak-38.

While vertical take-off jump jets may have had their day, 'thrust vectoring' as developed for the Harrier is a technique that will be used to provide a short take-off capability for the Harrier's replacement carrier-borne strike aircraft, the F-35 Lightning.

THE WORLD'S MOST DEVASTATING WEAPON

(ZYKLON B)

There's nothing about the name 'hydrocyanic acid' that would make any normal person want to go anywhere near the stuff. In fact, it's perfectly harmless if handled correctly and it was only horrific military inventiveness that turned it into the most appalling weapon of all time.

Hydrocyanic acid was used as a pesticide on orange crops in the United States and Spain in the late 19th century and was developed into a product that was invaluable for fumigating food stores, ships, railway freight cars and factories, ridding them of insect and rodent infestations. When sprayed in dilute form, it could even be used as a delousing treatment for humans.

By 1920, German chemists had developed hydrocyanic acid into a product called Zyklon A, a more stable form of the substance, with the scientists also having given it a particular 'warning' smell to avoid anyone accidentally inhaling the gas. Within a couple of years, Zyklon A had been transformed into Zyklon B, a form of the chemical that was far easier to handle, with the liquid having been absorbed into diatomite, used nowadays to make cat litter, and later still formed into crystals that released the Zyklon B as a gas when water was added.

Although attempts had been made to use hydrocyanic acid as a poison gas during the First World War, there was nothing much of interest to the military in Zyklon B except, perhaps, for delousing troops and their barrack rooms. That was all to change when SS Captain Karl Fritzsch turned his twisted mind to the subject.

Fritzsch was an ill-educated, lowly merchant seaman until he joined the Nazi Party at the age of 27 in 1930. He quickly found himself at home in the SS and by 1940 he was second-in-command at the Auschwitz-Birkenau concentration camp. Fritzsch had a reputation as a monster, condemning one group of prisoners whose friends had escaped to be locked in a basement until they starved to death as a warning to others against making escape attempts. It was Fritzsch, who may have had some previous experience with Zyklon B from his days working on boats on the Danube, who first experimented with the substance by locking prisoners in a basement and releasing the gas into the room. Up to 25 Soviet prisoners died in Fritzsch's initial experiments in 1941 but within weeks 600 Soviet prisoners of war (PoWs) and 250 sick Polish PoWs were subjected to the same treatment. Zyklon B was turned from a pesticide into a weapon of mass destruction and its production was ordered without the warning odour, making its use on a large scale less obvious. It was to claim the lives of an estimated 1.2 million prisoners.

Fritzsch eventually committed crimes that even the SS could not ignore. He was court martialled as a result of an investigation into corruption and sent to serve as a combat soldier. He then disappeared and, although some believe he was killed during the fall of Berlin, there is evidence that he escaped to Norway where he was captured by British Intelligence and left in the care of a Jewish guard …

THE PIRATE'S BIGGEST WORRY

(NAVAL MINES)

Piracy was the scourge of Chinese coastal trade in the Middle Ages with Japanese, Korean , Chinese and even Portugese bandits attacking ships and settlements, sometimes with the co-operation of corrupt local Chinese officials. These were no random bandit gangs, but organised armies of criminal cut-throats and the Chinese fought them with every weapon at their disposal. One such weapon was an early form of naval mine.

Wooden boxes or clay pots sealed with wax or putty and packed with gunpowder were set adrift by Chinese ships, judging the wind, the current and the length of fuse required to bring the floating mines alongside the pirate vessels before they exploded. The pirates, perhaps thinking the flotsam to be of value, might even have attempted to haul the mines aboard. They would soon learn the error of their ways. The charge that could be packed into a mine in this way would have been far more destructive than other long-range weapons available at the time.

The Chinese also developed mines that could be detonated remotely. The mines would be anchored at the mouth of a river or entrance to a port and, as an enemy ship passed within range, they could be detonated by someone hidden on shore pulling a cord that struck a flint on steel inside the mine, detonating the gunpowder.

Gunpowder, of course, does not work very well if it is damp and keeping anything that bobs about on the ocean completely dry is a real challenge, so such devices generally had a short lifespan during which they would not have been entirely reliable.

With the advent of modern explosives and industrial engineering in the 19th century, the naval mine became a hugely important strategic weapon. Land mines were most useful as defensive weapons, protecting an area from the approach of an enemy, but naval mines could be used both defensively and offensively. When German forces were poised in France during the Second World War, ready to invade Britain, they planned to defend their crossing of the English Channel by laying vast minefields to stop the Royal Navy getting close to their invasion fleet. When the time came for the Allies to invade France in 1944, they had to clear paths through German minefields protecting the coast.

Mines capable of being dropped by aircraft or laid by ships or even submarines have been used throughout the 20th century to blockade ports, to prevent enemy warships or merchant vessels from leaving, or to render vital waterways or trade routes too dangerous to use.

An entire array of mines has been developed over the years and modern mines come in many different forms. Some are anchored to the sea bed, bobbing on the surface; some are anchored but do not reach the surface, their cables setting them at depths where they can be of use against submarines; some lie on the bottom in wait for submarines; some can detect ships passing and release themselves from the bottom; and others can sense the presence of a ship and release a homing torpedo or even a rocket.

During the Second World War magnetic mines were used that were activated when a large metal object – a ship – passed nearby. They did not need to be in contact with the ship as the shock wave from their explosion was powerful enough travelling through the water to rip open a ship's plates. To counter such mines, minesweeper ships trailed long cables through which they passed an electric current that detonated the devices.

The most recognisable form of mine is the steel sphere that floated in the water, bristling with prongs. This is a contact mine and the bristles are called Herz horns. The first type of Herz horn was invented by Immanuel Nobel, the father of dynamite inventor Alfred Nobel, who placed a glass phial inside the horn, containing acid. When the horn was crushed on contact with a ship, the phial broke and the acid mixed with other chemicals to produce heat and flame that ignited the explosives. In later versions the acid ran into a battery, causing the battery to produce an electric charge that detonated the mine.

Hundreds of thousands of mines were laid during the Second World War and many of them are still out there, finding and clearing all of them having proved to be an impossible task. Old contact mines are especially dangerous because the simplicity of their design means that some of them are probably still capable of detonation.

Japanese pirates had far less to fear from the original mines, but they had their own worries. One of the punishments used by the Chinese as a way of executing captured pirates was to put them in a large cauldron and boil them alive!

OUTRUNNING
THEIR OWN NOISE

(SUPERSONIC AIRCRAFT)

Breaking the 'sound barrier' to fly supersonic represented a huge advance in aircraft technology and the race to go supersonic was, naturally, led by the military.

During the Second World War, aircraft became capable of very high speeds with the first operational jet fighter, the Messerschmitt 262 'Swallow', capable of around 560 mph and the Messerschmitt 163 'Komet' rocket plane achieving 600 mph. The Swallow was not produced in enough quantity to have any influence on the outcome of the war and the Komet was dangerously unreliable but they were both more than 100 mph faster than any Allied combat aircraft. At that time, straight line speed, especially when that 100 mph represented a 25 per cent difference in your speed, was paramount. Allied pilots could not catch the Swallow pilots to engage them, unless they were able to dive down on them, and the only way they could shoot them down was by trying to ambush them when they were either landing or taking off.

As the aircraft flew faster, however, they began to encounter a speed 'barrier'. Swallow pilots reported experiencing violent buffeting in high-speed dives and Allied pilots also experienced turbulence and loss of control under similar conditions. It seemed that, even though the engine had plenty more power to give, there was a speed beyond which it was not possible to fly. In fact, it was possible to fly faster, but there was a hurdle to overcome that became known as the 'sound barrier'.

When any aircraft flies across the sky it pushes air out of its way in the same way that a boat must push water out of its way when sailing on the sea. Just as the boat creates waves in the water in front, to the sides and behind it, so the aircraft creates waves in the air. These pressure waves travel through the air at the speed of sound. The speed of sound in air varies according to air pressure and temperature but at sea level and 20° Celsius it is about 761 mph. That's one mile every five seconds, which is why, when you see a flash of lightning, you can count the seconds until you hear the thunder then divide by five to work out roughly how far away the storm is.

When an aircraft begins to approach the speed of sound, it starts to catch up with the pressure waves it has created and the waves it is still creating crash into the ones ahead of them, creating turbulence in much the same way that waves converging in water will do. The aircraft then flies through that turbulence, which is the buffeting that pilots flying high-speed aircraft experienced.

The first aircraft to break the sound barrier was the Bell X-1 experimental rocket plane in 1946 and it paved the way for the development of aircraft that incorporated design features that would allow them to fly faster than sound without loss of control. Wings with leading edges that were swept back played a major part in controlling air flow, as did jet power, propellers creating all sorts of turbulence of their own that added to the buffeting problem.

Once jets were able to break the sound barrier, they literally left the buffeting problem behind, travelling faster than their own pressure waves and outrunning them. In overtaking the pressure waves, a supersonic aircraft causes them to merge into one and the resulting shock wave can be heard by observers on the ground as a sonic boom, like a clap of thunder.

Some military jets are now able to fly at many times the speed of sound but the only civilian airliners ever designed for supersonic flight were Concorde and the Soviet Concorde lookalike, the Tu-144, nicknamed 'Concordski'. The Soviet aircraft flew very few scheduled flights and was retired after only six years in service but Concorde was operated on scheduled flights from 1976 to 2003. The aircraft covered the distance between New York and Paris in 3.5 hours, less than half the time taken by other commercial aircraft.

For passenger flights, however, the advantages of time saved using supersonic flight against the cost involved meant that when Concorde came to the end of its working life it was not replaced by another supersonic airliner. It was far cheaper to carry more people on slower aircraft, although as technology advances, that situation is almost certain to change in the future.

THE ARMY MORNING WAKE-UP CALL

(BUGLE)

For any unfortunate young soldier enjoying the tail end of a particularly good dream after a sound night's sleep in an army barracks almost anywhere in the world, the dream ends with 'Reveille' being blasted out on a brass bugle.

The sleepy soldier, now wide awake, could be forgiven for thinking that the bugle was designed purely for an all-out assault on his eardrums – because it was. Sending signals using wind instruments has been a means of communication for centuries. The Romans used bugles made of animal horn as long ago as the 4th century but it is the brass bugle that really made its mark. Its strident tone can be heard even above the clamour of battle and that, rather than getting a bleary-eyed footslogger out of bed, is what made it such a valuable military invention.

Developed from musical instruments in the 18th century, the modern bugle has since changed very little. There are no sophisticated valves or sliders to produce notes on a bugle. It is a simple, spiral brass tube with a bell end for amplification. There are only five notes played in a bugle call and each sequence means something different – from 'time to get up' to 'time to eat' and 'forward charge'!

First used in Germany around 1758, the bugle proved an ideal way of communicating orders in barracks, on the parade ground or in battle. Its use spread throughout Europe and to the United States where, if you were a settler heading west whose wagon train was drawn into a defensive circle as you fought off warring natives or despicable bandits, the sweetest sound you could ever hope to hear was a US Cavalry bugle sounding the charge. On the other hand, the eeriest and most unnerving sound some American and British soldiers in Korea heard, given that by 1950 they were well used to using radio communications in the field, was the shrill sound of Chinese buglers. The Chinese did not have radios below regimental level, so battalion units continued to use bugles and runners.

First thing in the morning, however, there's still nothing that pops the sleep out of a squaddie's eyes than 'Reveille', which comes from the French for 'wake up'.

SMALL BIRDS
AND WIZARDS

(MERLIN ENGINE)

There's nothing like the sound of a PV-12 to set your pulse racing – any petrolhead will tell you that. When you then ask, 'A what?', stand by for one of those conversations involving crank cases, horse power, cylinder head and evaporative cooling systems … the sort of conversation that will dry out your eyeballs. Your alternative is to outwit the engine junkie by heading him off at the pass with, 'Ah, yes – the Merlin', and use that to change the subject to small predatory birds or wizards.

Just in case that doesn't work, you'll need to know a bit about the engine that played such a huge part in the Allied victory during the Second World War. The Merlin was so named because Rolls-Royce liked to use the names of birds of prey for their engines and this engine was to become the most famous the company every designed. It began life in the mid-1930s when Britain was desperately playing catch-up in an aviation arms race with Nazi Germany.

In May 1935 the Germans began testing a prototype monoplane fighter that was rumoured to be revolutionary. The Messerschmitt Bf 109 had a low wing, an enclosed cockpit, retractable landing gear and all-metal construction that put it light years ahead of the canvas-clad biplanes being flown by the rest of the world. It was also to have a powerful new Daimler-Benz engine, although the prototypes took to the air powered by specially acquired Rolls-Royce Kestrel units.

The British response came in the shape of two new aircraft, the Hawker Hurricane and the Supermarine Spitfire. The less sophisticated Hurricane first flew in November 1935, powered by an early version of the Merlin engine. Using tried-and-tested techniques, the Hurricane had a cloth-covered fuselage and wings but still performed far better than any other aircraft then in service with the RAF. In 1936, the all-metal Spitfire took to the air, demonstrating speed and aerobatic potential far in advance of the Hurricane. Both fighter aircraft were to go into production, the Hurricane entering service in 1937 and the first Spitfires being introduced in August 1938, by which time the Messerschmitt Bf 109 was already being tested in combat during the Spanish Civil War.

When battle was joined in the Second World War, the Bf 109 proved itself superior to the Hurricane in some respects but the argument still rages about whether the Messerschmitt was ever a better aircraft than the Spitfire. During the Battle of Britain, the Merlin-powered Hurricanes and Spitfires were the fighter aircraft that defended the United Kingdom, holding out tenaciously against the *Luftwaffe* onslaught until the Germans finally cut their losses and turned their attention to the invasion of the Soviet Union instead of the invasion of Britain.

Merlin engines were manufactured in factories in Derby, Crewe, Manchester and Glasgow as fast as they could be produced. The Rolls-Royce unit proved itself to be the best-performing aero engine of its day with most of those produced being installed not in fighter aircraft but in the four-engined Lancaster Bomber. The Merlin powered a range of planes that included the North American P-51 Mustang, perhaps the finest fighter aircraft of the war. The Mustang used a Packard engine which was a Rolls-Royce Merlin built under licence in America.

One of the most surprising aircraft to be powered by the Merlin was the Messerschmitt Bf 109. In service with the Spanish Air Force after the Second World War and manufactured under licence in Spain, some of the aircraft were fitted with surplus Merlin engines bought from Rolls-Royce. Merlin-powered Messerschmitts even flew against Spitfires and Hurricanes in the 1969 *Battle of Britain* movie.

Without having to know anything about gaskets or grommets, you should now be able to hold your own against the petrolhead.

FRANKLY, MY DEAR ...

(BROWNING .50 CAL)

When you have starred in the most successful film in box office history and are widely regarded as one of Hollywood's greatest actors, no one could blame you for hogging the limelight for as long as your fame lasted. Clark Gable was nominated for a Best Actor Oscar in 1940 for his role as Rhett Butler in *Gone With The Wind*, having previously been nominated twice and won in 1935 for his role in the 1934 movie *It Happened One Night*. You wouldn't dream of putting a career like that at risk. Clark Gable did. He gave it all up for a Browning .50 calibre machine gun.

By 1942, when Gable turned his back on Hollywood, although the MGM studio never quite let him go, using its influence with the US Army Air Force to try to make sure that their hottest property stayed in one piece, he did so to become a waist gunner in a B-17 Flying Fortress, with the mighty Browning in his hands to have a crack at the enemy.

The Browning .50 cal, officially designated the M2, was to become almost as famous as Gable. The US military had no home-grown heavy machine gun when they became involved in the First World War and the Browning evolved from designs created to remedy that. By 1933 the Browning M2 was ready for production in a number of different variants designed specifically either for infantry use, for anti-aircraft use, or to be mounted in an aircraft like Gable's gun. The aircraft version had the highest rate of fire at around 850 rounds per minute and the weapon had an effective range of well over a mile with its .50 inch diameter rounds doing considerable damage to all but heavily armoured targets.

The stopping power and reliability of the .50 cal led to it being used not only by gunners in heavy bombers but by fighter planes as well. Spitfires were initially fitted with four and then eight Brownings, as were Hurricanes during the Battle of Britain.

The Browning is still in use with British and American forces as well as being the primary NATO heavy machine gun, making it the longest-serving weapon of its type in history.

Clark Gable used his many times during the handful of bombing raids in which he participated, although his active service was curtailed when he was asked to put his talents to better use making a recruiting film for the army. It would be nice to report that he used his famous and, at the time due to its use of profanity, quite shocking line from *Gone With The Wind*, 'Frankly, my dear, I don't give a damn', but he was actually too old for combat duties in any case. The Browning .50 cal, it would seem, will never be too old.

THE COMING
OF THE INTERNET

(THE INTERNET)

Military requirements are responsible for making so much of what we take for granted in our everyday lives possible. Your mobile phone with its satellite connections and GPS technology owes its existence to the military. Whenever you use your home computer, you have the military to thank and, if you are using it to do a bit of online shopping, that's all down to military research as well. Email and the internet only exist because of technology developed by the United States Department of Defense.

The internet began life as ARPANET, the Advanced Research Projects Agency Network. The Advanced Research Projects Agency (ARPA) was established in 1958 to conduct research into new technologies and the formation of the agency was directly linked to the launch of the Soviet *Sputnik* satellite. When the Soviet Union managed to put a satellite into orbit before the United States, the American government immediately allocated massive funding to the education system to produce new scientists and engineers. Their aim was to accelerate the research and development of new technology for the space race. ARPA's initial work was concerned with space, missile defence systems and the monitoring of nuclear tests. It was an agency of the Department of Defense and, as DARPA – the D is for Defence – it still is.

ARPA scientists worked in laboratories at universities and institutions all over America and as the 1960s progressed, computers grew to become an essential tool of their trade. Some computer facilities were, however, better than others and it was difficult for them to share information because each institution might have computer terminals linked to its own computer, but linking one location's computer system to another was not possible. Giving scientists access to the best possible resources and allowing them to talk to each other via computer essentially led to the establishment of ARPANET.

Creating ARPANET involved ground-breaking work at a time when modern electronic computers were still in their infancy but by 1969 the network was ready to test. Four institutions were connected – the University of California, Los Angeles (UCLA); the Stanford Research Institute in Menlo Park, California; the University of California, Santa Barbara (UCSB); and the University of Utah. The first message that was sent on the ARPANET was a test sent between UCLA and Stanford that was intended to be the word 'login' but the 'gin' was lost when the system crashed, so the first message was actually 'lo'.

Despite such teething problems, the network was soon up and running between the first four participants and by spring 1970 ARPANET was connected to the east coast of the United States. The number of connections grew steadily with international links to London and Norway coming in 1973. The system proved that networking could work and was the forerunner of the modern internet, although ARPANET remained dedicated to work concerning the United States Department of Defense.

HOME-MADE MUNITIONS

(THE IED)

What is now termed an IED – Improvised Explosive Device – sounds like a modern invention but is actually as old as the most primitive gunpowder explosives. An IED is a home-made bomb that utilises whatever the bomber has to hand in order to make the most destructive device possible. It is something that bombers working in basements have been building ever since the advent of explosives.

This may include using munitions such as artillery shells or hand grenades but can also involve using chemical fertiliser to create an explosive charge. All that is really required to create an IED is an explosive charge, a detonator and a trigger mechanism. Such bombs can be extremely powerful, making them the favoured weapons of terrorists and those involved in combat or warfare that would be considered irregular or unconventional – guerrilla groups that do not have the manpower or conventional firepower to engage a well-equipped army openly.

Hidden by roadsides, in culverts or in buildings, IEDs can be triggered like a booby trap by tripwire or timer but are often detonated remotely by radio control or even by a signal from a mobile phone, allowing whoever planted the bomb complete control over precisely when to explode the device for the most devastating effect.

The IED has become the weapon that has claimed the majority of casualties and caused the most fatalities amongst Coalition soldiers fighting the Taliban in Afghanistan but they have also been used in Iraq, India, Northern Ireland and pretty much everywhere else in the world that non-regular fighters battle against regular military units.

While car bombs, suicide belts and parcel bombs all come under the catch-all term of IED, it is their use as roadside devices against security patrols that has brought the term into popular parlance.

DRESSING LIKE NODDY

(NBC SUIT)

No soldier in his right mind would ever want to dress like Noddy on the battlefield. The fictional boy created by children's writer Enid Blyton famously wore a bright red shirt, a red-and-yellow spotted scarf, unfeasibly large shoes and a pointed blue hat with a bell on the end. While something like that might do for the Swiss Guards at the Vatican, it's hardly disruptive pattern camouflage.

The 'Noddy' suit is actually the nickname given by British squaddies to their Nuclear, Biological and Chemical (NBC) warfare suits. These suits were first developed in the 1950s when, with the proliferation of nuclear and chemical weapons, it was feared that soldiers might have to fight in areas heavily contaminated by fallout or toxins. The suits were worn over a soldier's regular kit, with a helmet, gloves and boots worn over his combat boots to seal him in completely. Some British versions had a pointed helmet, which is why they came to be known as 'Noddy' suits.

Wearing the suit also meant breathing through a respirator, or using a completely separate air supply. The first suits were heavy and awkward, made from rubberized canvas that was impregnable to gas, toxins or radiation, but also held in the heat and sweat generated by a soldier struggling under the weight of his kit. They could only be worn for a certain amount of time before the soldier started to wilt from heat exhaustion.

Modern suits are made from lightweight reinforced nylon and lined with charcoal-impregnated felt that acts as a filter. They are more comfortable to wear because they allow heat and moisture to vent out of the suit, but they do not provide such a high level of protection and must be regularly replaced.

NBC suits, also known as CBRN (Chemical, Biological, Radiological and Nuclear) suits, were developed for the soldier in the field but have been adapted for civilian use by emergency services when dealing with road accidents involving chemical spillages or industrial accidents where hazardous material is on site.

Showing commendable even-handedness, British squaddies who first saw Soviet NBC suits, which had long, pointed face masks to house the respirator and incorporated round goggles, called them 'Wombles' after the furry creatures said to live on Wimbledon Common.

THE BREN ICON

(BREN GUN)

In 1935 the British Army was in the market for a new light machine gun and, after extensive trials, it selected a new design called the ZB vz.26 that was made by the Zbrojovka Brno Factory. The machine gun was to be produced at the Royal Small Arms Factory in Enfield, London and the names of Brno and Enfield were combined to rename the weapon the Bren.

The Bren became a familiar sight to soldiers throughout the rest of the 20th century, serving with all arms of the British forces as well as with Commonwealth forces during the Second World War, in Korea, the Suez Crisis, in Northern Ireland and even in the Falklands War in 1991.

Robust and reliable, the Bren used the same ammunition as a British soldier's Lee-Enfield .303 inch rifle and it was used as a 'squad' weapon to give each small detachment of soldiers some heavy-duty firepower. The distinctive ammunition pouches on the front of the Tommies' kit were designed to hold magazines for the Bren with every man expected to carry at least two. Every soldier was also trained to use the Bren should the need arise.

When the Lee-Enfield was replaced as the standard infantry weapon in 1957 with the L1A1 SLR, which used new 7.62mm

NATO ammunition, the Bren was adapted to take the new rounds rather than being retired. It was in this form that the Bren went into action during the Falklands campaign.

During the Second World War, the Bren made Veronica Foster famous as 'Ronnie the Bren Gun Girl' when she was photographed in Canadian munitions factory John Inglis & Co. Posing smoking a cigarette and admiring the Bren on which she had just finished working, Veronica became a propaganda poster girl in the same vein as America's Rosie the Riveter.

CREDIT WHERE IT'S DUE

(THE JERRYCAN)

Simple, clever, functional design is always to be admired. When something is properly thought out in a logical fashion and then manufactured to a good standard, it is almost certain to do the job for which it was intended and to go on doing so for a lifetime.

German engineers have a reputation for creating functional, logical products. Karl Daimler and Gottlieb Benz did precisely that with their engines and motor cars, yet there is still room for a little whimsy in the German psyche. The cuckoo clock is German, after all, not Swiss as so many people seem to think.

When it came to designing a military fuel can, therefore, the Germans built the best fuel can in the world, with a few novel features thrown in. The can did not have a screw cap as might be expected but a lid that employed a lever system and in the neck of the opening there was an integral spout for ease of pouring. The lid sealed closed with a gasket to avoid spillage. There were three handles arranged abreast of one another on top of the can, enabling a full can to be carried by two people, each using one hand on an outside handle; or to be carried by one person using the inside handle for better balance. The three handles also made it easier to pass cans down a line of men when they were being loaded onto vehicles or stacked for storage. Needless to say, they were designed to be stacked in regimented rows.

The handles were hollow and, when the can was full, they provided an air pocket to help cope with expansion should the contents become warm. An air pipe from the neck of the can allowed air to enter via the handles to help with smooth pouring.

The sides were indented with a cross-shaped corrugation to strengthen the steel structure and also to help with expansion, while the inside was lined with plastic, allowing the can to be used either for fuel or for water. The whole thing was about the size of a suitcase, holding 20 litres of liquid, and prior to the beginning of the Second World War, the German military stockpiled many thousands of these cans.

The British forces, on the other hand, transported fuel in thin metal boxes that flexed and wobbled so much that they were known as 'flimsies'. Pouring from one required the use of a funnel and it was a difficult task to accomplish without some degree of spillage. When the British came across the German Army fuel cans, officially termed the *Wehrmacht-Einheitskanister,* they adopted them for their own use whenever they could capture the cans. They also set about copying the design.

The origin of the design was, however, always acknowledged, the British giving credit to the Germans for inventing the can in the way that the British named it. British slang for 'German' during the Second World War was 'Jerry' and the cans were known as Jerrycans. Similar cans, still known as Jerrycans throughout the English-speaking world, continue to be used not only by military units but by campers, farmers, aid organisations and anyone else who needs to carry water or fuel with them when they venture out into the wilds.

PARKING DINKS AND AIR DEFENCE

(RADAR)

When you are backing your car into a parking space and it gives you a helpful beeping sound that gets faster and faster the closer you come to putting a dent in the car behind you, you are using a kind of radar, a technology that was developed in great secrecy for military use.

In the late 19th century, scientists were beginning to understand the nature of radio waves and the fact that the waves, or signals, could be reflected off certain surfaces. Metal objects were very good at reflecting radio waves and in 1904 German inventor Christian Hülsmeyer showed how the presence of a ship could be detected using radio waves without being able to see the ship itself, either at night or in fog. His system could not show direction or distance. By the 1920s, researchers in Britain, Germany, America, France, Japan and pretty much every developed nation in the world were working on ways to build a viable radio detection system.

In the early 1930s, engineers working at the US Naval Research Laboratory were able to send out a pulse radio signal that would reflect back to a receiver showing range to a target and the US Army developed a system that could be used to focus coastal searchlights on targets at sea.

A British team, working for the Air Ministry under Robert Watson-Watt, had originally been asked to look into using radio or energy beams as weapons, but quickly concluded that was not possible, although detecting aircraft was. They perfected a system that sent out radio signals which, when reflected off an aircraft, could be picked up by an aerial array and displayed on a screen. The system could identify the direction and distance to the target as well as its altitude, but it was far from flawless. When rushed into service in 1939, the Chain Home stations could detect aircraft as far away as 150 miles but struggled if the target was flying below 5,000 feet (1500 metres). Chain Home Low stations were built to pick up incoming aircraft down to 2,000 feet (610 metres) but even when further improvements were made, very low flying aircraft could go undetected.

Nevertheless, Britain had the world's first integrated air defence system, with radar stations around the coast reporting contacts to a central command from which interceptor aircraft could be directed to engage the enemy. Germany had its own version of radar, although it was not as well developed as the British system at the beginning of the Second World War and, when the Chain Home stations were set up, the German intelligence services were able to identify their purpose from the signals that the stations were broadcasting. Attempts were made to destroy the stations by bombing them, but the British rebuilt them so quickly that the Germans eventually gave up, assuming that because their system was not as advanced, the British would not be able to use theirs to any great effect.

While the British had the first working radar network, it was the Americans who gave it its name. RADAR was short for Radio Detection And Ranging. Developed out of military necessity prior to the Second World War, radar is now in use in every country in the world with air traffic controllers relying on it to keep airports operating safely, ships and aircraft using it for navigation, and drivers now relying on it to avoid fender benders!

MULTI-STOREY WARFARE

(SIEGE TOWERS)

In a battlefield situation, the average soldier wants to keep his head down, find an inconspicuous place to stay hidden and, maybe, fire his weapon at the enemy if he can manage it without everyone else using him for target practice. Going into battle in a multi-storey building hardly seems like the average soldier's version of a good idea.

If you were outside the walls of a city three thousand years ago, on the other hand, it was a brilliant idea, albeit something of a last resort. If you have found it impossible to break down the walls, or tunnel in, building a tower that allows you to see over the walls, fire arrows or other missiles into the city or even extend ladders so that you can storm the parapet is a reasonably logical plan.

Siege towers, as they became known, were first used in the Middle East around 900 BC and became ever more sophisticated as the years rolled by. A siege tower might take weeks to build and required a great deal of timber. For this reason it might be brought along to the battle in kit form, like other siege engines, and assembled on site.

Building the tower out of range of the city's defensive weapons – arrows or missiles hurled from a trebuchet – it would then be rolled forward to the wall on giant wheels. The wooden structure would be covered in thick animal hide soaked in water to try to prevent the defenders setting light to it with fire arrows, and on each level of the tower, archers or crossbowmen would fire at anyone who appeared on the walls to take pot shots at them.

Once you were close to the wall, if you were at the city gates you could deploy a battering ram from the base of your tower to try to smash the gates or simply slog it out with the defenders until your men could fight their way onto the city wall.

To counter the threat from siege engines, castles developed moats that made it impossible to bring a siege tower to the wall without first bridging the moat, although there was often time to do just that. The longest siege in English history took place at Kenilworth Castle in Warwickshire in 1266 when King Henry III spent six months outside the walls, plenty of time to create access to the walls for several giant siege towers each of which was armed with a number of trebuchets and up to 200 archers.

Even as late as the 16th century, siege towers were still in use, providing firing platforms for cannon that might otherwise not be able to lob their shot over the walls into the city.

THE BOMBER'S OLDEST ENEMY

(LASER-GUIDED BOMBS)

Anything that uses a laser to guide it onto its target must surely be the most accurate, most modern type of space-age bomb imaginable. In fact, the first laser-guided bombs were developed by the US Air Force as long ago as 1968.

The bombs were intended to be used to hit designated targets as a way of solving the problem that aircraft have had ever since they were first used as bombers – hitting the right target. Precision bombing always having been a difficult goal to achieve under combat conditions, the first laser-guided system fared little better. Dropped from a Phantom jet, which had a pilot and a weapons officer, the bomb would adjust its trajectory in flight to home in on a target that was being 'painted', or illuminated with a laser beam. The laser beam was projected from a hand-held device aimed from the cockpit by the weapons officer. Keeping the laser on target was obviously a problem but this first system, called BOLT-117, showed that laser guidance could be made to work far better than dropping 'unguided' bombs.

With the improvements in electronics and the miniaturisation of components, the on-board computing power of a bomb or missile improved immeasurably over the following few years, as did the laser projectors, and new tactics for the use of laser-guided bombs were adopted. Special forces troops on the ground are now able to identify a target and 'paint' it with a laser, keeping it in their sights while the bomb is dropped from an aircraft high above them that they may not even be able to see. The bomb then homes in on the target with such accuracy that some special forces operatives claim to have been able to send them in through doors or windows.

The main problem with laser-guided 'smart' bombs today is the same problem that has always dogged anyone tasked with precision bombing missions – the weather. Just as spotting a target obscured by cloud was a nightmare for bomber crews during the Second World War, illuminating a target successfully in bad weather can pose problems and, more importantly, getting the bomb to home in on the laser is not always possible if clouds come between the bomb and its target.

Even the smartest of weapons can still be defeated by good old Mother Nature.